ENERGY SECURITY

ENERGY SECURITY

ECONOMICS, POLITICS, STRATEGIES, AND IMPLICATIONS

CARLOS PASCUAL
JONATHAN ELKIND
editors

BROOKINGS INSTITUTION PRESS
Washington, D.C.

Library of Congress Cataloging-in-Publication data
Energy security : economics, politics, strategies, and implications / Carlos
Pascual and Jonathan Elkind, editors.
     p.   cm.
  Includes bibliographical references and index.
  ISBN 978-0-8157-6919-4 (pbk. : alk. paper)
  1. Energy policy. 2. Power resources. 3. Climate change. 4. International
cooperation.
  I. Pascual, Carlos. II. Elkind, Jonathan. III. Title.
  HD9502.A2E55115 2009
  333.79—dc22   2009041176

Digital printing

Printed on acid-free paper

Typeset in Sabon with Strayhorn display

Composition by Circle Graphics
Columbia, Maryland

Contents

Foreword

As cochair, with Senator Richard Lugar, of the Energy Security Initiative
Advisory Council, I am very pleased to present *Energy Security:
Economics, Politics, Strategies, and Implications* as our debut publication. Today's challenges and unfolding events underscore the timeliness
and relevance of this volume, which seeks to explicate the major issues
underlying the need for a new approach to energy security—geopolitical
tensions, energy interdependence, and climate change—by bringing
together thoughtful essays from scholars representing a cross-section of
Brookings's core research programs, each of whom has expertise in aspects
of the energy security dilemma. While each chapter can stand on its own
merits in addressing a critical issue—such as China's energy industry, the
role of energy in global governance, or the links between urban planning
and climate change—and while each represents the distinctive views of
the author, I believe that the examination of these issues in a comparative
framework is one of the strengths of this book.

The Energy Security Initiative (ESI) at Brookings was launched on the
premise that energy security will be significantly enhanced if solutions are
found that take into consideration the need to balance the geopolitical,
economic, and environmental implications of energy. It also recognizes
that, in so complex an arena, policy trade-offs will, inevitably, have to
be made. The initiative gains great strength from its ability to bring
together the broad multidisciplinary strengths of Brookings and the rich

scholarly and policy experience of its researchers. ESI is guided by the understanding that energy policy choices of the past have shaped the current economic, environmental, and strategic landscape in profound ways—and that the energy decisions that we make today will have no less profound an impact on the future in those areas.

A research community such as Brookings has an obligation to help bring about a better understanding of the complexities of energy by exploring the factors that influence or are influenced by our decisions. The chapters that follow will, in their combination, contribute to a deeper understanding of the choices that we have and the consequences of each and provide the analysis and insights required for making sound policy decisions.

DANIEL YERGIN
Chair, IHS Cambridge Energy
Research Associates

December 2009

Introduction

CARLOS PASCUAL and JONATHAN ELKIND

Energy is at the heart of economic development in every country. It moves us and powers our factories, government and office buildings, schools, and hospitals. It heats homes and keeps perishable foods cold. Its centrality explains its complexity. Energy is the source of wealth and competition, the basis of political controversy and technological innovation, and the core of an epochal challenge to our global environment. This book presents a collection of chapters on the theme of energy security. In this volume, the contributors seek to promote thoughtful analysis and healthy debate about different aspects of energy security through examination of the major factors currently driving energy policy decisions, including the actions of other nations, a changing climate, and the quest for energy independence. There is no unanimity here—no orthodoxy. Instead there are insights into different aspects of energy security and its relationship to the geopolitical, national, and environmental questions of our day.

The United States has been debating energy security since the oil crises of the 1970s, and indeed many of the solutions proposed during the most recent spike of oil prices could be mistaken for the solutions touted from previous decades—such as support for increasing the domestic supply of oil. But we are no longer in the 1970s; the world stage and the global energy landscape have both changed dramatically. Projected growth in the demand for energy from non-OECD countries such as China and India will exceed demand growth in the industrialized world. Economies are more integrally linked through globalization; thus we are more dependent on global trading partners for continued development. And the human race now confronts one of its greatest challenges—halting the threat of global climate change that results largely from the burning of fossil fuels.

Defining "Energy Security"

Running throughout this volume is the question of the proper definition of the term "energy security." This is no simple issue. The notion of energy security hinges on perspective: the temporal choices that we make and the way that we balance economic, national security, and environmental concerns. If energy security has ceased to be defined by the simple terms of affordability and dependable supply, what then do we mean when we refer to "energy security" today?

For some leaders and writers in the United States, energy security has come to be synonymous with "energy independence"; the two terms are now being used almost interchangeably in the political discourse. This view finds little support in the chapters that follow. On the contrary, our contributors question the wisdom—and even the practicability—of this goal. They subscribe to the view that our energy security will, for several coming decades, depend profoundly on petroleum and thus on secure international trade in energy.

One of the central points implied in the chapters that follow is the question of how we manage the transition from today's energy economy to the new, low-carbon energy economy that must be our future. In the short term, that implies the need to sustain, and indeed expand, existing relationships with our chief energy suppliers. At the same time, we must dramatically accelerate progress toward the technologies and trading patterns that we will need to meet our long-term goals of a significantly decarbonized energy market by 2050 or thereabouts. Several of the chapters present ideas for how best to execute this necessarily bifurcated strategy. The chapters are divided by subject into three parts, detailed below: "Geopolitics," "Understanding Energy Interdependence," and "Climate Change."

Geopolitics

The first part of this volume deals with the geopolitical aspects of energy security, which involve the management of the energy-related relationships that exist among states. Perhaps the most fundamental relationship is the one between energy suppliers and consumers, but important relationships also take shape between and among competing consumer countries or groups of consumers. There are also the dynamics that arise from the economic importance of energy and the risks posed by the interruption of supply links. Finally, there are relationships that move beyond mutuality of benefit, relationships in which one party or another seeks to exploit its energy-related power to dictate other aspects of political or security relations with another country.

All of these dynamics are experiencing additional stress in the current period because of the global financial crisis. As global energy demand drops, and with it global energy prices, producer countries are subjected to sudden and potentially severe fiscal strains. Moreover, global economic uncertainty is causing many energy companies to delay making investments, opening the possibility of impending energy shortages once the global economy starts to rebound.

In chapter 1, "Geopolitics of Energy: From Security to Survival," Carlos Pascual and Evie Zambetakis survey the links between current and future energy relationships on one hand and security concerns on the other. They assess how energy security has gone beyond the geopolitical relations among states, affecting both the risks of climate change and nuclear proliferation, issues that threaten global as well as national security.

In "Energy Security in the Persian Gulf: Opportunities and Challenges," chapter 2, Suzanne Maloney highlights key recent developments in the Gulf, that most essential petroleum production region, finding the possibility of a virtuous cycle from revenue flows where in the past there has been instability and limited developmental impact.

Michael O'Hanlon examines another aspect of our dependence on oil and gas production from the Persian Gulf—the cost of military forces to protect the global energy trade—in chapter 3, "How Much Does the United States Spend Protecting Persian Gulf Oil?"

Chapter 4, "Who's Afraid of China's Oil Companies?" is the final chapter in this part of the volume. In recent years, China has emerged not only as a major energy importer, but also as a major commercial player in international markets. In analyzing the motivation and goals of Chinese energy companies, Erica Downs finds reason to believe that Chinese energy companies behave more like other international enterprises than some analysts have been prepared to admit.

Understanding Energy Interdependence

If the contributors in this volume view "energy independence" as a distant or even illusory objective, there still remains the question of how countries can most effectively manage their energy *inter*dependence—their energy relationships. Individual countries routinely choose certain actions and forgo others that define the extent and the impact of their energy dependence.

Some countries rely extensively on international energy trade—energy dependence. Producer countries require energy demand in order to monetize their natural endowment and create the wealth that can contribute to social welfare if that wealth is managed effectively. Producers therefore may tend to seek higher prices as long as those prices do not induce demand destruction—fuel switching, technological advances, and general reduction in energy consumption. Consumer countries rely on abundant supply—or at least smooth-functioning energy markets—to ensure adequate supplies of energy for their economies.

Consumer countries may give strong emphasis to improving efficiency and reducing energy intensity to limit the degree to which their economies rely on energy production and imports, seeking thereby to insulate themselves from some of the vicissitudes of the global energy market. Important questions remain, however, as to whether and when this policy approach is economically rational. No less important are questions about whether and how producer and consumer countries can engage in productive dialogue to contribute to mutually beneficial improvements in governance.

In chapter 5, "Making Sense of 'Energy Independence,' " Pietro Nivola and Erin Carter examine data related to energy independence and the behavior of a number of international actors. They administer a "reality check" and reject the idea that avoiding international trade in energy yields beneficial, sustainable policy.

Jonathan Elkind calls for an updated definition of energy security in chapter 6, "Energy Security: Call for a Broader Agenda." He points out that as the United States and other leading nations now try to intensify their response to the challenge of a warming climate, it is essential to emphasize efficiency and ensure that actions meant to enhance the availability, reliability, and affordability of energy supplies do not work at cross purposes with our environmental objectives.

The final contribution in this part of the book is chapter 7, Ann Florini's "Global Governance and Energy." Florini examines the difficulty of creating more effective international institutions that can enhance stable and sustainable international energy relationships.

Climate Change

Climate change is arguably the greatest challenge facing the human race. It poses profound risks to the natural systems that sustain life on Earth and consequently creates great challenges for human lives, national economies, nations' security, and international governance. New scientific reports emerging from one year to the next detail ever more alarming potential impacts and risks.

It is increasingly common for analysts and policymakers to refer to climate change as a threat multiplier, a destructive force that will exacerbate existing social, environmental, economic, and humanitarian stresses. The warming climate is predicted to bring about prolonged droughts in already dry regions, flooding along coasts and even inland rivers, an overall increase in severe weather events, rising seas, and the spread of disease, to cite just a few examples. Such impacts may spark conflict in weak states, lead to the displacement of millions of people, create environmental refugees, and intensify competition over increasingly scarce resources.

One of the great challenges of climate change is, indeed, the scope of the phenomenon. The ongoing warming of the globe results chiefly from one of the most ubiquitous of human practices, the conversion of fossil fuels into energy through simple combustion. Halting and reversing climate change, however, will require both unproven—perhaps even unimagined—technology and *sustained* political commitment. We must change living habits in all corners of the globe over the course of the next several decades. We must resist the impulse to leave the problem for those who follow us or to relax our efforts if we achieve a few years of promising progress. The profound challenge will lie in the need for successive rounds of sustained policymaking, successive waves of technological innovation, and ongoing evolution of the ways in which we live our lives.

Marilyn A. Brown, Frank Southworth, and Andrea Sarzynski tackle another critical aspect of our response to climate change in chapter 8, "Features of Climate-Smart Metropolitan Economies." Urbanization has been occurring in the United States, as elsewhere around the globe, for decades, and this trend has great significance for climate change. For the first time in history, more people now live in urban than in rural areas. Cities are the drivers of economic growth, but they are also responsible for the production of some 70 to 80 percent of global emissions. Choices made in regard to the specific nature of U.S. metropolitan areas—for example, policies to encourage high-density development and the use of public transportation—therefore can have a large impact on the level of greenhouse gas emissions.

Jason Bordoff, Manasi Deshpande, and Pascal Noel address policy options for climate change mitigation in chapter 9, "Understanding the Interaction between Energy Security and Climate Change Policy." They focus in particular on policy instruments that can help to reduce energy-related GHG emissions while also promoting traditional notions of energy security.

In chapter 10, "Five 'Gs': Lessons from World Trade for Governing Global Climate," William Antholis speaks to the difficulty of creating the institutional structures needed to allow the world to negotiate and implement an international climate agreement. Existing institutions, from the United Nations to the G-8, seem ill-suited or insufficiently structured to face the task. Our ability to succeed in developing a proper global response may require drawing lessons in governance from the global trade regime and the creation of the World Trade Organization.

The challenge of energy security has been the topic of decades-long debate because it goes to the heart of each nation's economy and so many existing relationships around the globe. Policy options for addressing energy security are highly consequential for our economic and security prospects. Even more complex, a person's individual perspective colors significantly the extent to which he or she perceives risks to energy security and opportunities to enhance it. If the person is from a petroleum-producing country or a coal-producing state, he may favor a prescription that is very different from that favored by someone who lives in an energy-efficient urban region or a low-lying island state. The Brookings Institution offers these ten chapters as a contribution to what promises to be a lively debate and a defining issue for our times.

PART I

Geopolitics

CHAPTER ONE

The Geopolitics of Energy
From Security to Survival

CARLOS PASCUAL and EVIE ZAMBETAKIS

Since the industrial revolution, the geopolitics of energy—who supplies and reliably secures energy at affordable prices—has been a driver of global prosperity and security. Over the coming decades, energy politics will determine the survival of life as we know it on our planet.

The political aspect of energy, linked to the sources of supply and demand, comes to public attention at moments of crisis. When unstable oil markets drive up prices and volatility hinders long-run investment planning, politicians hear their constituents protest. But energy politics have become yet more complex. Transportation systems, particularly in the United States, are largely reliant on oil, so disruption of oil markets can bring a great power to a standstill. Access to energy is critical to sustaining growth in China and India—to employ the hundreds of millions who remain poor and to keep pace with burgeoning populations. Failure to deliver on the hope of greater prosperity could unravel even authoritarian regimes—and even more so democratic ones—as populations become more educated and demanding.

Two of the major global energy consumers, the United States and the European Union, have similar needs but different practical perspectives on energy imports. The United States is overly dependent and focused on oil, with consequent special attention to the Middle East. The EU is highly reliant on imported gas, making Russia an important supplier and factor in the EU's energy policies and raising tensions particularly between

Germany and the central European states. Before the onset of the 2008 financial crisis, rising demand for oil and gas imports and limited capacity to expand short-term supply drove up prices, supplier wealth, and producer leverage, allowing producers such as Russia, Venezuela and Iran to punch above their weight in regional and international politics. With the current slowdown in global demand from at least the traditional demand centers in Europe and the United States, lower oil prices have rattled the economies and politics of producer states that have come to depend on large export revenues to maintain stability at home and support muscular foreign policies abroad. That is especially poignant in countries like Iran and Venezuela, which highly subsidize social programs and fuel at the expense of economic growth and diversification.

Traditional geopolitical considerations have become even more complex with global climate change. The U.N. Intergovernmental Panel on Climate Change (IPCC) has documented that the use of fossil fuels is the principal cause of increases in atmospheric concentrations of greenhouse gases, which in turn are driving up the mean temperature of the planet. A changing global climate is already resulting in significant loss of glaciers and shrinkage of polar icepacks. It will lead to severe flooding in some places and drought in others, which will devastate many countries' food production, encourage the spread of various illnesses, and cause hundreds of thousands of deaths each year, particularly for those living in the developing world. Nearly 2 billion people were affected by weather-related disasters in the 1990s, and that rate may double in the next decade.[1] At the same time as countries are competing for energy, they must radically change how they use and conserve energy. The politics of the debate over scrambling to secure hydrocarbon resources versus reducing consumption through efficiency and use of alternatives—particularly how to pay for the cost and dissemination of new technologies and how to compensate those who contribute little to climate change but will experience its most severe effects—is emerging as a new focal point in the geopolitics of energy.

Ironically, volatile oil and gas prices and the actions that must be taken to address climate change—namely, pricing carbon at a cost that will drive investment, new technology, and conservation to control its emission— will drive another existential threat: the risk of nuclear proliferation. Higher energy and carbon prices will make nuclear power a more attractive option in national energy strategies, and the more reliant that countries become on nuclear power, the more they will want to control the fuel cycle. The risk of breakout from civilian uses of nuclear power to weaponization will increase dramatically, as will the risk of materials and technology getting into the hands of terrorists.

Confronting these challenges requires an understanding of the fragility of international oil and gas markets and also of the nexus among energy security, climate change, and nuclear energy and proliferation. This chapter addresses that interconnection and the kinds of measures that will be needed to ensure a politically, economically, and environmentally sustainable energy strategy.

Shallow Markets, Sharp Politics

International economic and political developments can exacerbate the effects of inelastic supply and demand on global energy markets, causing massive price fluctuations even when the underlying nature of the market remains unchanged. Under such volatile conditions, political power has accrued in the hands of energy exporters, making it more difficult to gain consensus among net importers on international policies, such as deploying international peacekeeping forces to Darfur and imposing sanctions on Iran to gain leverage against the risk of nuclear weaponization. And price volatility has also exacerbated the impact of bad economic policies in energy-exporting states when revenues have collapsed during economic downturns—dealing a critical blow in the collapse of the Soviet state in 1991, for example.[2] Over the long term, reducing market volatility serves the self-interest of both energy importers and exporters.

To frame this discussion, recall that the price of oil rose from $21 a barrel at the beginning of 2002 in the run-up to the Iraq war, to $29 at the start of hostilities on March 19, 2003, to $48 at the start of President Bush's second term in January 2005, to $145 in July 2008[3]—an overall rise of over 400 percent. Prices then fell during the recession in late 2008, hovering at about $50 a barrel in the spring of 2009 with decreased consumer demand.[4]

To change the dynamics of energy markets from instability to security, both importers and exporters must get beyond the cyclical price incentives that perpetuate the current structure of international oil and gas markets. For net importers, that will mean diversifying energy sources, with greater reliance on renewable energy and energy conservation. For exporters, that will mean internal economic diversification to reduce dependence on export revenues. Yet when energy prices are high, exporters have generally used revenues to consume more. When energy prices are low, the

FIGURE 1-1. Oil Demand and Supply Balance, 1970–2008

Barrels per day, millions

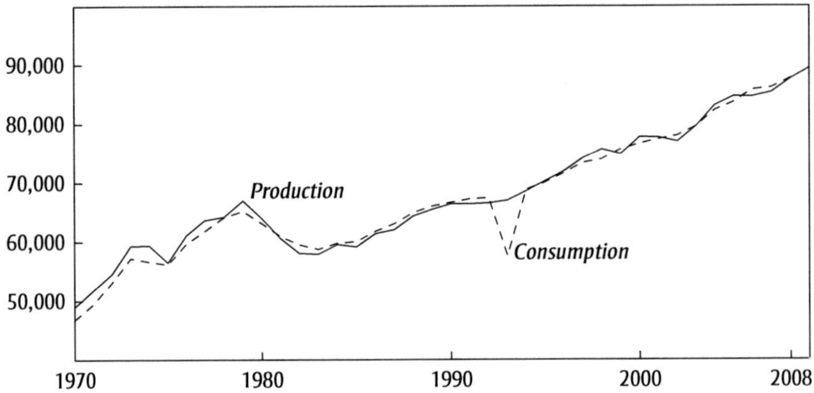

Source: Energy Information Administration, *International Petroleum Monthly* (February 2008).

political will to tax energy to create incentives for conservation and inno-vation sharply diminish. The result, illustrated in figure 1-1, has been an almost tandem rise of international oil production and consumption, with the exception of a sharp drop in consumption in 1992–93 when the Soviet Union collapsed. Until political leaders break this mismatch in pricing and political incentives, the underlying structure of oil and gas markets will continue to undermine the long-term security interests of both importers and exporters.

Figure 1-2 illustrates the demand and supply factors behind oil price volatility. Bloc 1 in the chart represents the fastest-growing sources of demand for oil: the United States and China. Bloc 2 consists of Saudi Ara-bia, Russia, Iran, Iraq, Venezuela, Nigeria, and Kazakhstan. These are countries upon which oil importers de facto rely to meet short-term sup-ply shortages. Bloc 3—Canada, the United Kingdom, Brazil, India, Japan, Norway, and Indonesia—shows other important drivers of supply or demand, most notably Japan and India, which rely massively on oil imports.

On the supply side, there is limited ability to expand production rapidly in the short term, and even long-term prospects are mixed. Figure 1-3 shows that in the past decade, Russia and Saudi Arabia have accounted for the largest increases in oil supply. Existing Russian fields are now pro-

FIGURE 1-2. Oil Production and Consumption, 2007

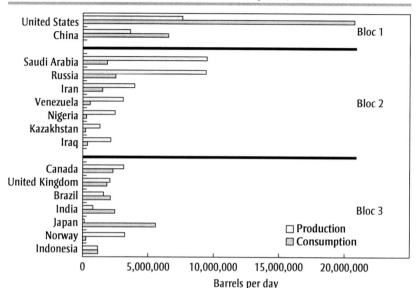

Source: See CIA *World Factbook* 2007 and 2008 (www.cia.gov/library/publications/the-world-factbook/); figures through third quarter of 2007.

ducing at their peak, and Saudi Arabia has limited additional short-term capacity. Due to commercial disputes, local instability, or ideology, Russia, Venezuela, Iran, Nigeria, Mexico, and Iraq are not investing in new long-term production capacity.[5]

Given limited supply elasticity, political volatility gets magnified through fluctuating and unpredictable prices. Key sources of instability include conflict in the Middle East, the risk of the Iraq war spilling into the Persian Gulf, the risk of U.S. and/or Israeli conflict with Iran over its nuclear program or over Iranian support for militias in Iraq, conflict in the Niger Delta, populist state controls in Iran and Venezuela, and the difficulty of securing major oil transport routes. Saudi Arabia pledged to increase oil production by 200,000 barrels a day of heavy sour crude at the Jeddah Summit on June 22, 2008, which was essentially offset by off-shore attacks on Shell's $3.6 billion "Bonga" floating production, storage, and offloading vessel on June 19 by the Movement for the Emancipation of the Niger Delta (MEND), which, in combination with kidnappings of oil workers and sabotage of onshore pipeline infrastructure, kept between

FIGURE 1-3. Petroleum Production, 1996–2008

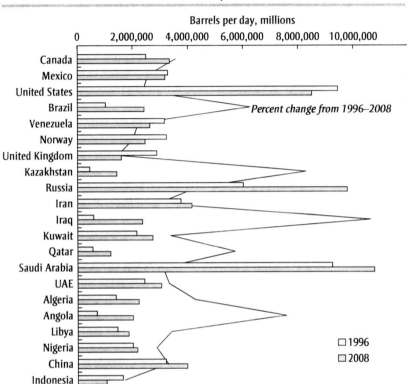

Source: See Energy Information Administration, "International Energy Statistics" (http://tonto.eia.doe.gov/cfapps/ipdbproject/IEDIndex3.cfm?tid=5&pid=53&aid=1).

600,000 and 900,000 barrels a day of Nigerian high-quality crude output off-line. Despite efforts to repair infrastructure, Nigeria—once Africa's largest oil producer—is, under these circumstances, being outpaced by Angola and branded an unreliable producer, thus underscoring the limits of energy security in a tight supply environment.

Political risk is exacerbated by choke points in transit routes. Nearly 40 percent of world oil exports pass through the Strait of Hormuz, nearly 28 percent through the Strait of Malacca, and nearly 7 percent through Bab el-Mandeb, the narrow strait connecting the Red Sea and the

Gulf of Aden.[6] Tehran's threats in 2007 to block the Strait of Hormuz if attacked over its nuclear program illustrates how several energy issues— oil transit, civilian nuclear energy use, and nuclear proliferation—can be intertwined in a volatile mix of international security and conflict. The difficulty of getting pirate attacks around the Horn of Africa under control, if they had occurred in 2008 rather than 2009, could have had disastrous impacts on energy prices when prices were already soaring. Yet in the context of a global recession in 2009, the price impact has been limited.

Supply-side fragility is accompanied by limited elasticity of oil demand in the short run, a result of the transportation sector's high level of reliance on gasoline and other petroleum-based motor fuels. Figure 1-4 illustrates how the United States and China have driven the largest share of rising oil demand since the mid-1990s. Change in this arena, such as switching to alternative fuels, requires long-term investments in technology and infrastructure. In the medium term, there are options such as increased use of hybrid cars that plug into the electricity grid.[7] Ironically, the 2009 recession could further entrench the structural factors that could cause a return to increased demand for oil in both the United States and China. In the United States, a temporary spike in demand for hybrid vehicles in the summer of 2008 turned into an about 30 percent year-on-year reduction in demand in January 2009.[8] That, together with the overall crippling of the auto industry, which has driven Chrysler to bankruptcy, has made it harder for automakers to finance the transition of their fleets. Beyond that, economic pressures to create jobs quickly will drive economic stimulus funds toward infrastructure investments, and those investments that can be made most quickly are based on highway transit.

Against those structural factors, the massive price swings seen from peak oil prices of $145 a barrel in the summer of 2008 to about $50 a barrel in the spring of 2009 are easier to understand, even if the precise inflection points in price trends are hard to predict. First, the subprime mortgage crisis drove investors from real estate to oil and other commodities. Speculative oil demand exacerbated tight and costly supply, pushing oil prices upward. When the U.S. financial crisis turned into a global economic recession by late 2008, the demand and price trends reversed. The International Monetary Fund estimates that global GDP will contract by 1.3 percent in 2009, affecting both industrialized and emerging economies. Demand for energy has contracted with global GDP, as has speculative investment in energy commodities. U.S. crude oil consumption is down by 1.45 million barrels a day, which is 6.8 percent less than last year, and crude stocks

FIGURE 1-4. Petroleum Consumption, 1996–2007

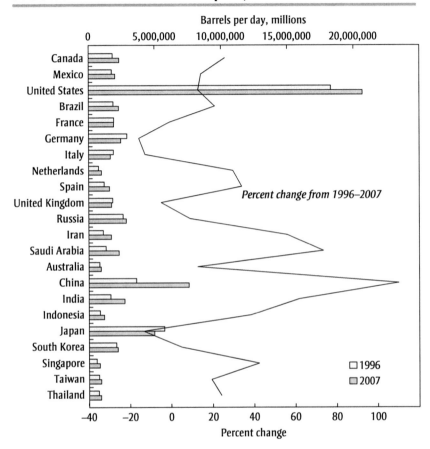

Source: See Energy Information Administration, "International Energy Statistics, 1996–2007" (http://tonto.eia. doe.gov/cfapps/ipdbproject/iedindex3.cfm?tid=5&pid=54&aid=2&cid=&syid=1996&eyid=2007&unit=TBPD).

rose by 5 million barrels in December 2008, which is the largest gain since 1970.[9] The result has been a reverse free fall down the price curve that brought energy to record highs in mid-2008.

Still, structural factors will likely drive an eventual price reversal. Falling prices have begun to curtail long-run investment in exploration and production (E&P) as more expensive projects are put on hold;[10] that, in turn, will feed back into the long-run outlook. E&P planned under high oil prices to bring online more oil and gas to alleviate the tight supply market will not have taken place on the size and scale needed. While some international oil companies claim that they will stick to their investment plans, OPEC indicates that about thirty-five new projects could be on hold, cutting by about half the increases projected in global production capacity expected by 2014.[11] As argued above, the recession constrains the capacity of the private sector to invest in massive restructuring in the short term to accelerate the transition to a less fossil fuel–intensive infrastructure base.

To get out of this cycle of volatility, then, national leaders will need to change the structure of energy markets and reduce dependence on both fossil fuels and fuel exports as a revenue source. That will require investments in conservation to reduce demand and to expand renewable sources of energy. Sustaining such investments will require consistent price signals to industry, investors, and consumers. And that will require national leaders to take actions that may have short-term financial and political costs. In the meantime, one of the costs paid is in U.S. national security due to the volatility to which we subject the economy and the power we transfer to energy suppliers willing to use their wealth in ways that complicate U.S. national interests.

Energy and Power Politics: Iran, Venezuela, and Russia

Iran, Venezuela, and Russia have had some of the most obvious political impacts on the realities of today's oil market. Their customers and investors have at times set aside their political concerns to preserve their commercial interests. All three countries have used their energy wealth and leverage to strengthen their regional influence with more vulnerable neighbors, and all three have used the stature that they have acquired through their regional interventions and wealth to complicate U.S. interests.

Iran is developing a nuclear program despite UN Security Council resolutions 1696, 1737, and 1747 demanding that Iran suspend the enrichment of uranium and fully discloses the nature of its nuclear program. When the International Atomic Energy Agency (IAEA) board of directors referred Iran to the UN Security Council (UNSC), countries from every part of the world opposed Iran's development of the capability to produce a nuclear weapon. Yet the country remains defiant.

In part, that may be out of the hope that Russia and China will block any serious sanctions, largely because of their commercial interests in Iran. China is moving into gas development projects in Iran, where Western companies are kept out by the sanctions regime. Both Russia and China have generally resisted using international sanctions to exert pressure on other countries, in part to serve their own commercial interests, in part to avoid precedents authorizing the UN to scrutinize sovereign decisions on national security. The National Intelligence Estimate (NIE) of 2007 found that Iran's nuclear weapons program had been suspended in 2003 and that it had not been restarted as of mid-2007. However, with indigenous civilian nuclear capacity and technical expertise, there is potential for breakout—although it is important to distinguish between aspirations for breakout and the ability to do so, given that building uranium enrichment and/or reprocessing capacity is far more complex than building a civilian nuclear reactor.

While high oil revenues do not translate directly into market power and influence for Iran, they can embolden the country's most militant leaders to assert themselves on the nuclear issue. With the recent fall in oil and gas prices, the same leaders are faced with the prospect of not being able to provide the massive fuel and social subsidies that buy support for their regimes. However, the global nature of the economic downturn could actually make it easier for President Mahmoud Ahmedinejad to pass painful subsidy reforms without squandering as much political capital in the process. Here, price volatility translates into political volatility.

President Hugo Chávez's engagement with China and Russia, which is based on the promise and ability to deliver on energy agreements in the future, is risky, considering that Venezuela cannot guarantee its capacity to meet future production projections. The difficulty of and costs involved in extraction of reserves and lack of adequate maintenance and investment in technology, infrastructure, and new drilling render Venezuela ill-equipped to meet and sustain current OPEC quotas. Chávez has done such damage to the investment climate that exploration and production have not risen with growing demand and higher oil prices. He is undermining the very industry on which the entire country's economy and welfare system is predicated, then looking to China and Russia to fill the void of foreign investment, while trying to gain political leverage by posturing against the United States.

In addition to being one of the world's top-ten oil producers and a top supplier to the United States, Venezuela's Orinoco tar sands are estimated to be the largest deposits of their kind in the world, potentially rivaling conventional world oil reserves. Their strategic importance for global energy is enhanced by improvements in extraction technology and by potential future recovery rates with the turn to unconventional oil. When oil prices recover from the financial downturn, of the unconventional sources for oil—including Canadian tar sands—the Orinoco tar sands are the most economical. A poor investment climate combined with aggressive political rhetoric, unsound economic policy, and the current economic crisis poses a risk for development of these reserves, which could enhance global oil supply.

Venezuela's influence must be seen in the wider context of globalization and its impact in Latin America. Globalization has helped millions in Latin America to tap into technology, markets, and capital in a way that has made many countries and people wealthier. However, the gap between the "haves" and "have nots" has grown. Those who have not made it are increasingly better-educated and resentful for what they do not have. That resentment is strongest among those who are making the transition out of poverty but who cannot see how to advance further. Such individuals become vulnerable to populism, and when given a chance to vote, many will use their ballots to express their frustration. It is in this context that Venezuela and Hugo Chávez have brought their wealth to bear. Chávez's message of populism and his support for local leaders have the potential to galvanize local frustrations within Brazil and Mexico. In Bolivia and Nicaragua, the Chávez myth, seen from the outside, suggests that the poor could be given more at little cost.

Not every Latin American country has gone down Chávez's populist route, but he presents new challenges to a regional order based on democracy and market principles. For democrats in the region, the first challenge is to ensure that there is not a backlash against democracy from those leaders and countries that feel threatened by popular frustration. The second is to reform governance and policies to give the "have nots" the sense that they can have a better future. Whether Latin American leaders can educate their people to create the capacity to benefit from globalization, whether governments can target subsidies to those who need to be pulled into society, and whether the United States will open its markets to technologies, services, and products—these factors together will fundamentally affect perceptions of democratization in the region and whether it becomes a source of stability or a vent for populism.

Russia's veto power in the UN Security Council; its unique position in supplying gas, electricity, and oil to Europe; and its control over one of the two largest nuclear arsenals in the world make it important to understand the ways in which energy has transformed Russia internally and the nature of its role in the international community. In addition to being the world's

TABLE 1-1. Total Energy Consumption

	Gas consumption 2006	Gas imports from Russia 2006	Russian imports as a percent of 2006 gas consumption
Slovakia	5.5	6.30	114.44
Finland	4.3	4.52	106.08
Bulgaria	3.0	2.85	93.82
Czech Republic	8.5	7.13	84.06
Greece	3.2	2.40	74.04
Austria	9.4	6.85	72.89
Hungary	12.5	8.32	66.45
Turkey	30.5	19.65	64.44
Poland	13.7	7.00	51.12
Germany	87.2	36.54	41.91
Italy	77.1	22.92	29.73
Romania	17.0	3.95	23.26
France	45.2	9.50	21.04
Switzerland	3.0	0.37	12.42

Source: "BP Statistical Review of World Energy, 2007 and 2008" (www.bp.com/sectiongenericarticle.do?category Id=9023783&contentId=7044475) and Energy Information Administration, "Russia Country Analysis Brief" (www.eia.doe.gov/cabs/Russia/NaturalGas.html). Full-year data for 2006 are the most current published data as of this writing. Countries that import more than 100 percent of their gas consumption are either using the excess volumes to replenish national gas stores or are re-exporting a portion of their imports.

second-largest exporter of oil, Russia has the world's largest proven gas reserves—it controls over a quarter of the world's reserves, or 47,040 billion cubic meters—and also has the world's largest electricity grid.

Table 1-1 illustrates the importance of Russia's role as gas supplier for Europe. On average, European countries rely on Russia for 23 percent of their imported gas (the equivalent of three-quarters of Russian gas exports), and that number is expected to grow (depending on what happens with new Norwegian Arctic gas discoveries, which are expected to double current production levels from a dwindling North Sea supply). Russia's dominance in the primary energy mix is much higher among a number of eastern and central European countries. In this sense, Russian gas supplies can determine the economic vitality of Germany, Greece, Austria, Finland, and others. Generally, pipeline gas connections tend to create a long-term mutual dependence that militates against confrontational acts such as cut-offs or boycotts by the producer, the consumer, or the transmitter. Thus, even at the height of the cold war, gas supplies from the USSR to central and Western Europe continued without interruptions. However, in the last decade Russia has repeatedly demonstrated its willingness to use gas as a political weapon, in conjunction with commercial arguments about price, most vividly during confrontations with Ukraine in January 2006 and February-March 2008.

Oil is a fungible commodity, whereas natural gas delivered by pipeline—as most of the world's natural gas is, despite the nascent growth of a potential global market in liquefied natural gas (LNG)—entails a more concrete relationship between a discrete producer and a discrete set of consumers. Diversification of gas supply therefore is costly and requires a time-consuming licensing and construction process. New infrastructure, in turn, requires contractual commitments to underwrite financing for what are often multibillion-dollar projects. For example, the Nord Stream gas pipeline—known previously as the North European Gas Pipeline (NEGP)[12]—will connect gas fields in the Khanty-Mansiysk Autonomous Oblast to German and other European consumers. Two parallel pipelines will be laid under the Baltic Sea from near Vyborg in Russia to near Greifswald in Germany, with a capacity of 27 billion cubic meters a year for each of the two "threads." The first thread is meant to be commissioned in 2010 and the second in 2012. Assuming that Gazprom's plans proceed as announced, Nord Stream will have the capacity to deliver nearly 25 percent of Europe's incremental gas import needs by 2015. However, many industry experts think that Nord Stream will experience construction delays and that its ultimate cost will be a multiple of the initial price tag of €5 billion ($7.68 billion).[13] Nord Stream, moreover, will only further entrench Germany's dependence on Russian gas.

Russia's energy market power has allowed Russia to consolidate political power internally and has made Russia resistant to external political influence. Within Russia, former president Putin reversed the halting trend toward democratization that occurred through the 1990s by controlling the appointments of governors and the upper house of parliament and consolidating control over most broadcast media. He orchestrated a change in rules for parties to get into the lower house of parliament, in turn tightening the ties between political parties and the Kremlin. He appointed Kremlin officials to corporate leadership positions in the gas, oil, rail, airline, shipping, diamond, nuclear fuel, and telecommunications industries.[14] With power thus centralized, Putin rejected in increasingly belligerent tones any external criticism of Russia's political system and policy choices. He accused the Organization for Security and Cooperation in Europe of aiming "to deprive the [December 2007 parliamentary] elections of legitimacy" by pulling out of plans to observe them.[15] Russia continues to refuse to ratify the Energy Charter Treaty, which would set the terms for energy production and transit in Russia and other countries. Despite virtually every country in the word rejecting Russia's decision to recognize the "independence" that it orchestrated for South Ossetia and Abkhazia after its incursion into Georgia in August 2008, Russia has been immune to external pressures to relent on its position.

It is in this context that the United States and Russia now purport to hit a "reset" button on their relationship. Russia's policies toward Iran and whether it cooperates with the United States and the rest of the international community to avert Iran's acquisition of a nuclear weapon will be the most significant test of whether Russia believes that its energy wealth allows it to ignore wider accountability for its actions.

On one hand, Russia has stated that it has no interest in having Iran acquire nuclear weapons, and it has been part of the group of the five permanent Security Council members and Germany that is involved in negotiations with Iran. At the same time, Russia has resisted the imposition of tough sanctions against Iran, seeking to carve out exceptions for Russia's sale of civilian nuclear technology for Iran's Bushehr nuclear power plant and to weaken UN sanctions while providing cover for China to follow suit. Russian officials or former officials have indicated that they see prospects for the International Atomic Energy Agency to close out the file concerning the historical questions about Iran's nuclear program. According to these individuals, that will require returning the Iran case from the UNSC to the IAEA.

Russia, in effect, has positioned itself either to unravel or make viable an effective diplomatic package against Iran. If it splits the "P5 plus 1" (the five permanent UNSC members plus Germany) by insisting that the UNSC should not consider sanctions against Iran, Russia will undermine any effective diplomatic effort, giving Iran further leeway and virtually ensuring that it develops nuclear weapons capability. Such actions will raise the risk of U.S., Israeli, or other military action against Iran. Yet Russia also has the capacity to make clear to Iran—and to its Islamic constituents and neighbors—that the international community is not blocking Iran from a civilian nuclear program. To the contrary, Russia's cooperation can make it possible to offer Iran a more advanced civilian nuclear plant, assurances of enriched uranium fuel, and provisions for transfer of spent fuel back to Russia.

The Iran case and Russia's role in it underscore key elements of today's complex geopolitics of energy: market power to act in isolation, leveraging energy power through veto power at the UN, emerging risks and opportunities associated with civilian nuclear power, structural dependence embedded in gas markets and pipelines, and limited recourse to use international rules to promote accountability. For consumer nations—and those who see the wider risks of vesting so much political power in energy-rich states—the short-term options are limited, as production is managed by producer countries. Better management of consumers' emergency inventories could help, and bringing China and India into an emergency stocks management system would seem crucial since they are the biggest drivers of increased oil demand yet are outside the International Energy Association's stocks management system. The more critical changes come in the medium term, through conservation, alternative fuels, massive lifestyle changes, new building codes, and new technologies that burn less energy. It is these very types of policies that are also central to a different yet even more existential aspect of the geopolitics of energy: climate change.

The Geopolitics of Climate Change

Avoiding the destruction of the planet through the emission of greenhouse gases (GHGs) is one of the most complex challenges that the human race has ever created. Climate change puts the survival of many natural systems and biodiversity at stake, potentially leading to a myriad of deleterious consequences for human security. The difficulties lie in the intersection of earth sciences, technology, economics, and politics. The emission of greenhouse gases will have the same impact regardless of the source—Beijing, Detroit, or Newcastle—hence it is impossible to solve the global problem without involving all states or at least the major GHG emitters. The problem of human-induced climate change arising from the concentration of greenhouse gases in the atmosphere was created by the industrialized world, so emerging market economies resent that they must share the cost of avoiding or responding to the problem. Yet emerging economies are the fastest-growing source of greenhouse gas emissions. Deforestation accounts for 20 to 25 percent, which is roughly equivalent to U.S. emissions.[16] Worse yet, the biggest catastrophic impacts will be on developing countries, such as Mali and Bangladesh, that are not driving the problem.[17]

Science, technology, and domestic politics further complicate the picture and split even the developed economies. Figure 1-5, from the Intergovernmental Panel on Climate Change, illustrates the interrelationships among temperature, GHG concentrations, and impacts of a changing

FIGURE 1-5. Examples of Impacts Associated with Global Average Temperature Change[a]

Global average annual temperature change relative to 1980–99 (°C)

	0	1	2	3	4	5

WATER
- Increased water availablity in moist tropics and high latitudes – – – – – – – – – – – – – – – – –▶
- Decreasing water availability and increasing drought in mid-latitudes and semiarid low latitudes – – – – –▶
- Hundreds of millions of people exposed to increased water stress – – – – – – – – – – – – – – – – –▶

ECOSYSTEMS
- ——————— Up to 30% of species at ———————— Significant[b] extinctions ——▶
  increasing risk of extinction   around the globe
- Increased coral bleaching —— Most corals bleached —— Widespread coral mortality– – – – – – – – – – –▶
- Terrestrial biosphere tends toward a net carbon source as:
  ~15% ——————————————————— ~40% of ecosystems affected ——▶
- Increasing species range shifts and wildfire risks
- Ecosystem changes caused by weakening of the meridional – – – –▶
  overturning circulation

FOOD
- Complex, localized negative impacts on small-holders, subsistence farmers, and fishers   – – – – – – – –▶
- Tendencies for cereal productivity ——————— Productivity of all cereals – – – –▶
  to decrease in low latitudes   decreases in low latitudes
- Tendencies for some cereal productivity ——————— Cereal productivity to
  to increase in mid-to-high latitudes   decrease in some regions

COASTS
- Increased damage from floods and storms –▶
- About 30% of
  global coastal – – – – – – – – – –▶
  wetlands lost[c]
- Millions more people could experience – – – – – – – – – – – –▶
  coastal flooding each year

HEALTH
- Increasing burden from malnutrition, diarrheal, cardiorespiratory, and infectious diseases – – – – –▶
- Increased morbidity and mortality from heat waves, floods, and droughts – – – – – – – – – – – –▶
- Changed distribution of some disease vectors –▶
- Substantial burden on health services – – – – –▶

	0	1	2	3	4	5

Warming by 2090–99 relative to 1980–99 for nonmitigation scenarios

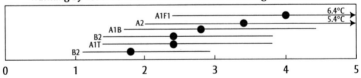

	0	1	2	3	4	5

Source: From Intergovernmental Panel on Climate Change, "Climate Change 2007: Synthesis Report—Summary for Policymakers."
a. Impacts will vary by extent of adaptation, rate of temperature change, and socioeconomic pathway.
b. Significant is defined here as more than 40 percent.
c. Based on average rate of sea level rise fo 4.2 millimeters/year from 2000 to 2080.

climate. The IPCC has stated that the maximum temperature increase that the world can sustain without suffering irreparable damage is about 2.0 degrees centigrade by 2050. There is less certainty about what concentration of GHGs will prevent anything more than a 2.0 degree temperature increase, but the estimates fall in the range of 450–550 parts per million (ppm) of CO_2e (equivalent carbon dioxide).[18] The lower the level, the costlier and harder it is to achieve. The world is currently at a level of about 420 ppm of CO_2e. There is also uncertainty about the level of annual reductions in greenhouse gas emissions that are needed to stabilize the atmosphere at a concentration of 450–550 ppm of CO_2e, but estimates range from 50 to 85 percent in annual reductions of CO_2e emissions relative to 1990 levels.

The objective of a climate change policy is to create the incentives that will drive changes in technology, technology dissemination, and consumption patterns and lead to new developments in how energy is produced in order to reduce the annual emission of carbon to a level that does not exceed 450 to 550 ppm by 2050. That is a monumental task. For example, if current practices and technology stay the same, estimates indicate that greenhouse gas emissions could increase by 25 to 90 percent by 2030 instead of decreasing on the order of 50 percent or more annually by 2050, which should be the trajectory.

Currently the technologies and policies to achieve that target do not exist. Conservation, efficiency, alternative fuels, and cleaner use of fuels all have to be part of the equation. However, the combinations currently available do not achieve the desired end. In order to succeed, the international community must find a way to price carbon in order to curb consumption, spur technological innovation, affect fuel choices, and stimulate investment. Some argue that, in the long term, there must be a stable long-term price for carbon of at least $30 per ton of CO_2e to achieve the necessary economic and technological incentives.

Yet pricing carbon has divided the world geopolitically. No country has adopted an explicit tax on carbon on the scale of $30 per ton. Cap-and-trade systems in Europe or those emerging in regions of the United States do not yet come close to that level of implicit carbon price. Within the United States, the more proactive states have adopted standards for the use of renewable fuels and fuel efficiency. Some states, like Florida and California, have set targets for overall GHG emissions, creating an implicit cost for carbon, but they are not setting the stable, explicit price signals that are needed for innovation. Japan, for example, has called for a 50 percent annual reduction in CO_2 emissions by 2050, but the Japanese government has kept a cap-and-trade system and a carbon tax off the table as policy options.

In addition, agreement not to subsidize domestic energy prices is a necessary component of any emissions control policy. Major energy producer and consumer nations alike distort domestic demand by subsidizing fuels. While India, China, and the producing states of the Middle East have recently begun to raise domestic energy prices, they continue to subsidize prices below their real cost of production; in contrast, if domestic consumers paid world market prices for petroleum and electricity, that would not only temper domestic demand but encourage efficiency improvements.

From the debates over policy, economics, technology and science during the Bush and now the Obama administration, four geopolitical blocs on climate change have emerged, with a fifth waiting in the wings. The first is anchored by Europe and to a lesser extent Japan, with both supporting the adoption of binding emissions targets. The second is driven by the United States together with Australia and supports setting a long-term goal with nationally binding medium-term commitments but not an internationally binding treaty that holds countries collectively to account. The third consists of the emerging market economies, led by China and India; it has resisted any form of binding international targets, focusing its demands on technology dissemination and financing for the cost differential for clean technologies. The fourth group comprises developing countries that will bear the brunt of flooding, desertification, and other catastrophic effects of climate change; their demands center on financing to adapt to the impacts of climate change. The emerging fifth group consists of energy suppliers who see the world shifting away from the use of fossil fuels. They could emerge either as facilitators of a transition toward a more carbon-free world if they invest their wealth in technology dissemination—and thereby position themselves as winners in a greener international environment—or they could act as spoilers, seeking to drive up prices and profits to capture the greatest earnings during the transition away from fossil fuels.

Among these groups, the United States has the capacity to play a pivotal role. China and India will not move toward more proactive domestic policies if the United States does not set the example. Along with Europe and Japan, the United States has the capacity to demonstrate that green technology and conservation can be compatible with growth and a foreign policy that is more independent of energy suppliers. The United States also stands to benefit from accelerated commercialization of green technologies and the development of global markets in energy-efficient and clean energy technologies. The ability of the United States to lead, however, will depend on domestic action—on whether it will undertake on a national basis a systematic strategy to price carbon and curb emissions. If it does, the scale and importance of the U.S. market can be a driver for global change. If it fails to act, then the United States will find that over time the opportunity for leadership to curb climate change will be replaced by the need for crisis management as localized wars, migration, poverty, and humanitarian catastrophes increasingly absorb international attention and resources. Eventually, its failure to act will come back to U.S. borders in a way that will make the Katrina disaster seem relatively tame.

The Geopolitics of Nuclear Proliferation

Perhaps the most existential risk, which parallels that of climate change, is that of nuclear technology and materials getting into the hands of rogue states or terrorist organizations. That could result in the devastation of cities or nations and set off reciprocal actions leading to the levels of destruction seen in Hiroshima and Nagasaki. High fossil fuel prices, the risks associated with energy suppliers and transport routes, and, ironically, policies to combat climate change—namely, the pricing of carbon— could accelerate the drive for civilian nuclear power, which could increase the risk. For economic, environmental, and security reasons, more and more countries can be expected to incorporate nuclear power into the mix of their power generation capabilities.

Today, just twelve of the fifty-six states with civilian research reactors— thirty of which have civilian nuclear power for electricity generation—can enrich and commercially produce uranium.[19] Arguably, nine countries currently have nuclear weapons: China, France, India, Israel, North Korea, Pakistan, Russia, United Kingdom, and United States. Most of these countries acquired nuclear weapons after acquiring civilian nuclear power capabilities (see figure 1-6). Nuclear weapon states have enough fissile material in their stockpiles to create tens of thousands of nuclear weapons, and there is enough separated plutonium (Pu-239) from civilian use to make just as many weapons. India diverted the plutonium used in its first nuclear test in 1974 from its Cirus research reactor a decade earlier. Imagine the risk if the number of nations producing enriched uranium were to double or triple as developing nations sought to enhance their

FIGURE 1-6. Nuclear Power States

Number of states

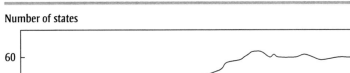

Source: Scott D. Sagan, "Nuclear Power without Nuclear Proliferation?" *Bulletin of the American Academy of Arts and Sciences* LXI, no. 2 (Winter 2008), p. 43.

energy security through a misguided sense of energy self-reliance while adopting carbon-free nuclear technology to produce electricity. That calls for an intensified effort now, before it is a crisis, to strengthen the fire-walls between civilian nuclear power and nuclear weapons programs.

A guaranteed external supply or "bank" of low-enriched uranium (LEU)—which can then be calibrated according to individual light-water reactor specifications (the most common type of reactor in use)—can serve as a back-up or reserve mechanism within the context of the existing global nuclear fuel market and should be sufficient if the real motivation is electricity generation for energy-starved states. As long as countries are fulfilling nonproliferation obligations, they should have access to LEU for nuclear fuel; according to IAEA director general Mohamed El-Baradei, that does not mean that states should give up their rights under the Nuclear Nonproliferation Treaty (NPT). Relevant proposals include the following:

—Global Nuclear Energy Partnership (GNEP): a U.S. proposal for the United States and international partners to supply developing countries with reliable access to nuclear fuel and emissions-free power generation in exchange for their commitment not to develop uranium enrichment and plutonium reprocessing technologies, thereby closing the fuel cycle

—Global Nuclear Power Infrastructure (GNPI): a Russian proposal for the creation of a system of international centers that will provide nuclear fuel cycle services under the supervision of the IAEA on a nondiscriminatory basis

—Nuclear Threat Initiative: a proposal to stockpile low-enriched uranium under the auspices of the IAEA as a last-resort fuel reserve for countries electing to forgo a national enrichment program.

The G-8 energy ministers acknowledge that nuclear nonproliferation and security should be ensured through agreed frameworks and international initiatives, such as GNEP and GNPI, in cooperation with international institutions such as the IAEA.[20] Regional entities such as the EU, NATO, ASEAN, and others also have engaged in nonproliferation activities and commitments.

Two major concerns, however, are that a world nuclear fuel bank could trigger a race in which states rush to join the nuclear club in the period before the bank is established and that an external bank could be perceived as an infringement on national sovereignty, with the result that economic incentives may not outweigh national or political imperatives. The Atoms for Peace program arguably facilitated India's and Pakistan's transition from peaceful nuclear technologies to nuclear weapons, while the NPT has been circumvented by the United States–India Peaceful Atomic Energy Cooperation Act. The potential for non-nuclear states to feel excluded and vulnerable needs to be addressed and mitigated.[21] The goal must be to give aspirants for civilian nuclear power the confidence to obtain nuclear fuel through an international fuel bank and to forgo enrichment programs while placing their entire nuclear programs under the IAEA Additional Protocol.[22] Furthermore, the World Bank and international financial institutions could finance nuclear plant construction as part of the deal for nuclear aspirants ratifying the Additional Protocol.[23] From the nonproliferation standpoint, it is better that a country import its centrifuges rather than develop the technology on its own. Such measures may not stop Iran's nuclear ambitions, but they may help other countries from breaking out from civilian nuclear programs to weaponization.[24] They will also reduce the risk of having nuclear material leak into the hands of rogue states and terrorists. To achieve the credibility necessary to lead the international community in forging such a revitalized regime against proliferation, the United States will need to follow through on the promises that it has made to what the non-nuclear weapons states see as "horizontal proliferation," namely ratification of the Comprehensive Test Ban Treaty (CTBT).

Realizing a safer international nuclear regime will require revitalizing the bargain between nuclear and non-nuclear weapons states under the Nuclear Nonproliferation Treaty. Article 4 of the NPT assures non-nuclear weapons states of their right to peaceful civilian application of nuclear power and to "the fullest possible exchange of nuclear technology" if they adhere to the treaty's provisions and forgo the pursuit of nuclear weapons. Since the drafting of the NPT in 1968, experience has demonstrated ways in which monitoring and surveillance should be enhanced to reduce the risk of leakage, and these measures have been incorporated into a voluntary Additional Protocol. In return, nuclear weapons states are committed under the NPT to reduce their arsenals and seek eventual nuclear disarmament.

It is the disarmament part of this agenda that former secretaries of state Henry Kissinger and George Shultz and former secretary of defense William Perry, along with former senator Sam Nunn, have proposed in their renewed call for the elimination of nuclear weapons.[25] Even those who think that full nuclear disarmament is unworkable or unwise recognize that U.S. ratification of the CTBT is the most critical step to restore the credibility and vitality of the bargain the NPT established between vertical (deepening within nuclear states) and horizontal (across states or other entities) proliferation. At the 1995 NPT review conference, non-nuclear weapons states accepted U.S. commitment to the ratification of the CTBT as a basis for the indefinite extension of the NPT—in effect, a deal for their permanent commitment to forgo nuclear weapons. In order to advance the actions needed now to curtail the vertical proliferation of nuclear weapons, the United States cannot ignore its 1995 commitment on CTBT.

A new package is needed on proliferation and testing that includes the following:

—a commitment by NPT signatories to accept the Additional Protocol
—development of an international fuel bank under the IAEA that would assure nations of a supply of nuclear fuel as long as they observe the NPT
—a means to centralize the control and storage of spent nuclear fuel
—a ban on testing that would complicate the ability of any aspirant for nuclear weapons to break out of a civilian nuclear program.

The ban on testing is pivotal in the geopolitics of nuclear power. A comprehensive test ban would have the greatest impact on states that want to use civilian programs as a platform for the development of nuclear weapons. Nuclear weapons states have other means to service and replenish their arsenals. Those truly committed to civilian nuclear power should not have a need to enrich uranium, and in most cases the scale would be sufficiently small that it would not make economic sense for them to do so. If any entity were to test a nuclear weapon, it should be immediately detectable, and it should trigger sharp multilateral pressure to abandon the program. That was the case with North Korea, when China, the United States, and Japan quickly secured UN condemnation and sanctions after North Korea's nuclear test in October 2006.

A comprehensive test ban creates the incentive to sustain the status quo among nuclear states and to constrain states from developing nuclear weapons capacity. The CTBT isolates those who seek to advance their ambitions for nuclear weapons. Russia would need to be part of this package—as a supplier of fuel and a secure source for storage and reprocessing—which would entail massive commercial benefits to Russia. The United States should seize on this opportunity to ratify and implement the CTBT and in so doing strengthen U.S. leverage to broker an international package to stop nuclear leakage and curtail the risk of breakout from civilian programs.

Conclusion

For more than a century, energy, politics and power have been clearly intertwined as a force in international security. The stakes are only getting bigger as the issues go beyond national prosperity and security to the viability of the planet. Policymakers and citizens must understand the nature of this change and recognize that inaction—simply not attempting to forge coalitions or provide constructive guidance on how states use energy—will be catastrophic.

It will be crucial to resist allowing short-term electoral cycles in the United States or elsewhere to drive energy policy and politics. Inevitably, some politicians will call for energy independence, an unrealistic and unattainable goal. That is simply not possible in an interconnected world that requires access to global markets, capital, and technology, whether a nation is a net importer or exporter of energy.

In the short term, diplomacy and effective reserve management will be critical tools, but they are not fully developed. Expansion of the International Energy Agency's reserve management system to China and India has failed several times for political reasons. Technical support to help China and India coordinate with others will be an important confidence-building measure because the two nations currently see themselves as pitted against the rest of the international community. Energy diplomacy also needs to be made a central foreign policy consideration. Key questions include the following:

—Where can nations jointly benefit from further exploration and development?

—What transit systems merit international cooperation and investment?

—Can regional security arrangements mitigate risk and create shared incentives across states, especially in the Middle East, the Persian Gulf, and Central Asia?

—Can the five permanent members of the UNSC reach an understanding to suspend the use of their veto rights on issues related to energy politics in order to stimulate a full debate around tough questions that get sidetracked through veto threats?

—Should nations commit to an E-15 group, composed of the largest economies and energy users, as a means to force a focus and sustained agenda on the policies and politics behind energy supply and use?

—How do domestic energy and economic growth concerns drive the foreign policy choices of China and India and their roles in multilateral institutions?

Focused answers to those questions could be the foundation for national, regional, and international energy strategies that foster cooperation on energy issues rather than allow short-term political considerations to shape what generally may appear to be zero-sum competitive outcomes.

In the medium and long term, both geopolitical interests and environmental sustainability call for a radical departure from current patterns in the use of fossil fuels, which compromises the national security of most states and threatens the entire planet. A shared medium-term strategy among states to foster convergence on political, environmental, energy, and economic goals should include

—measures to price carbon emissions and to coordinate prices across states, if not create transnational carbon markets

—financing and policy measures to support the development, testing, demonstration, commercialization, and dissemination of clean and efficient technologies that can transform the terms of debate on energy use and climate change (for example, addressing liabilities associated with carbon capture and sequestration)

—means to stimulate investment in clean technologies to reduce private sector and temporal risk for the developed countries, to finance the differential between clean and traditional technologies for emerging economies, and to develop infrastructure and adapt to climatic changes in developing countries

—common international standards for firms to disclose the use of carbon and establish guidelines for emissions per unit value of output in order to promote public accountability and guide investment decisions

—a new form of an international framework for climate change that reflects the complexity of the interaction of technology, economics, and politics and leads to better and tighter standards for performance over time.

On the nuclear side, no issue is more important than creating a strong firewall between civilian power and weaponization programs now, before more countries seek to break out from civilian programs. Hard as that may be, it will be easier than getting new entrants into the ranks of nuclear weapons states to disarm. For this process to begin, the United States must start with ratification of the Comprehensive Test Ban Treaty, with India and Pakistan acting in concert with the United States.

These are major challenges, but they are not unattainable. If such actions are taken now, there is a chance that the geopolitics of energy can move the international community toward constructive long-term outcomes. If not, the geopolitics of energy will make all nations less secure and bring into question the very viability of their future.

Notes

1. Ian Noble, "Adaptation to Climate Change and the World Bank," presentation at UNFCCC seminar "Development and Transfer of Environmentally Sound Technology for Adaptation to Climate Change," Tobago, June 14–16, 2005.

2. Yegor Gaidar, *Collapse of an Empire: Lessons for Modern Russia* (Brookings, 2007), p. 39–70.

3. See Energy Information Administration, "Petroleum Navigator" (http://tonto.eia.doe.gov/dnav/pet/hist/rclc1d.htm).

4. Falling demand was due to the falling dollar and U.S. economic downturn.

5. Although that could change in Iraq with the invitation of oil companies, although the short-term nature of the contracts on offer might actually hinder efficient and rapid growth.

6. See Energy Information Administration (EIA), "World Oil Transit Choke-points" (www.eia.doe.gov/cabs/World_Oil_Transit_Chokepoints/Background.html).

7. David Sandalow, *Freedom from Oil: How the Next President Can End the United States' Oil Addiction* (New York: McGraw-Hill, 2008).

8. See www.mixedpower.com/toyota-prius/recession-hurts-sales-of-hybrid-cars-temporarily/.

9. See Energy Information Administration, "International Energy Statistics for Consumption and Stocks, 1970–2008" (http://tonto.eia.doe.gov/cfapps/ipdbproject/iedindex3.cfm?tid=5&pid=5&aid=5&cid=&syid=1970&eyid=2008&unit=MBBL; http://tonto.eia.doe.gov/cfapps/ipdbproject/IEDIndex3.cfm?tid=5&pid=54&aid=2).

10. See Adam Schreck, "IMF: Mideast Growth to Slide in 2009," for the impact of the recession on oil-producing countries in 2009 (www.google.com/hostednews/ap/article/ALeqM5hBx3VzsoP571dwg1s91L4t9C-lewD983BIVO0).

11. See "Falling Oil Supply Risks a Price Rise," March 27, 2009 (http://online.wsj.com/article/SB123808291973348921.html).

12. Originally the North European Gas Pipeline (NEGP), it was re-named Nord Stream gas pipeline in October 2006.

13. As of March 31, 2008, the estimated cost had risen to €7.4 billion (US$11.7 billion) (Interfax report).

14. Neil Buckley and Arkady Ostrovsky, "Putin's Allies Turn Russia into a Corporate State," *Financial Times,* June 18, 2006.

15. Oleg Shchedrov, "Putin Accuses U.S. of Meddling in Russian Vote," Reuters, November 26, 2007.

16. Valerie Kapos, Peter Herkenrath, and Lera Miles, "Reducing Emissions from Deforestation: A Key Opportunity for Attaining Multiple Benefits," Feburary 2007 (www.unep-wcmc.org/resources/publications/unep_wcmc%20RED%20Feb07.pdf).

17. One of the most densely populated and poorest nations in the world, Bangladesh's devastation of the largest mangrove forest in the world to make room for grazing animals and to harvest firewood is at least a force multiplier of the monsoon flooding exacerbated by climate change.

18. CO_2e signifies "equivalent carbon dioxide," which is the internationally recognized measurement of greenhouse emissions.

19. Brazil, China, France, Germany, India, Iran, Japan, Netherlands, Pakistan, Russia, United Kingdom, and United States.

20. From the Joint Statement by Energy Ministers of the G-8, the People's Republic of China, India, and the Republic of Korea, Aomori, Japan, on June 8, 2008.

21. A Fissile Material Cut-off Treaty (FMCT), originally proposed by the UN in the early 1990s, would oblige nuclear weapons states to refrain from producing fissile material for weapons. In 2004, President George W. Bush opposed international verification of an FMCT. Verification would be effective and technically feasible, and assurances could be made so that highly enriched uranium declared for naval reactor fuel would not be diverted to weapon use. "Global Fissile Material Report 2006: First Report of the International Panel on Fissile Materials" (www.fissilematerials.org), p. 2.

22. Reprocessing is not a critical part of the fuel cycle and can be postponed indefinitely by storing spent fuel. Reprocessing nuclear fuel is attractive to those who focus on supply scarcity, because it eliminates most high-level nuclear spent fuel waste—although it produces a much larger volume of intermediate- and low-level waste—and fuel scarcity concerns; hence, the push for national enrichment plants. However, repro-

cessing creates Pu-239, which carries a proliferation risk. The Additional Protocol strengthens and expands the IAEA's verification safeguards to ensure that non-nuclear state parties to the Nuclear Nonproliferation Treaty (NPT) are using nuclear materials only for peaceful purposes. The United States signed the Additional Protocol in 1998, but it is not in force because the necessary legislation to implement it has not yet been passed.

23. Debate currently surrounds the World Bank's $US5 million carbon fund and whether it should include a provision for nuclear financing.

24. In addition, more proliferation-resistant technologies, such as thorium reactors, should be promoted as part of an overall policy; however, the abundance of uranium supplies in geographically diverse locations will make creating incentives for such technology difficult.

25. George P. Shultz, William J. Perry, Henry A. Kissinger, and Sam Nunn, "A World Free of Nuclear Weapons," *Wall Street Journal*, January 4, 2007, p. A 15.

CHAPTER TWO

Energy Security in the Persian Gulf
Opportunities and Challenges

SUZANNE MALONEY

Two small news items juxtaposed on a single page of a Qatari newspaper in February 2008 offer a trenchant synopsis of the opportunities and challenges facing Middle Eastern states and by extension U.S. involvement in the region. One article details efforts to settle a series of strikes by expatriate laborers in Bahrain with an offer to raise salaries by $40 per month. Situated just below is a second piece, which details the record-setting price—$14 million—paid for a vanity license plate at a charity auction in the United Arab Emirates (UAE).[1] Between those extremes lie the promise and the peril of the Persian Gulf in the twenty-first century, a place of epic wealth and persistent risk factors, both of which have been profoundly intensified by a seven-year stretch of record oil prices and the subsequent crash of the global economy.

Between January 2002 and July 2008, the price of a barrel of oil on the world market increased from $18.68 to a record $145 per barrel. For the Gulf states, the rapid price escalation created an incredible windfall—at least $1.5 trillion between 2002 and 2006, which represents a doubling of profits over the previous five-year period.[2] Unlike all previous oil booms, the recent escalation was not the product of a supply disruption but rather the seemingly insatiable demand increases associated with the rapid development of the economies of China, India, and other Asian countries.

Since mid-2008, the images of wanton luxury in the Gulf have largely been supplanted by equally staggering reports of a precipitous free fall among the region's oil-rich societies as a result of the global economic crisis. Instead of multimillion-dollar license plates, the Gulf is now the scene of mounting deficits, plummeting state revenues, and fleeing expatriates abandoning their property and vehicles—all the aftereffects of a dramatic collapse of at least 70 percent in oil prices. Dubai, the symbol of the glittering new Gulf, has been hit especially hard, as its interconnections to the broader global economy and its reliance on an overvalued real estate market made it especially vulnerable to the forces at work over the past few years. Still, even in a time of global contraction, the basic logic of the boom still applies: a vast and growing Asia requires reliable sources of energy, and both geology and proximity favor the Gulf as the supplier of choice. As a result, even as the global economy stumbles, the Gulf's oil revenues will continue to provide a steady income stream that will slowly begin to rise again over the medium term.

For the region's boosters, the windfalls launched a new "golden age" in the Persian Gulf and broader Middle East.[3] Regional real estate and financial markets boomed; grandiose infrastructure projects once again began rising in the desert; expatriate capital, institutions, and executives were lured to the glittering city-states of the Gulf. Furthermore, the boom appeared to fuel more than just more conspicuous consumption, as the petro-states embraced the lessons of previous eras and used their increased resources to expand badly needed savings and investment, engage in serious economic reform programs, and repatriate more of their capital. The net effect, from the perspective of many regional leaders and investors, is that a region often perceived as mired in tradition and dangerously insulated from the globalization that has taken place elsewhere over recent decades began to embrace modernity and the old verities of Arab politics began to be supplanted by a competitive new technocracy. Proponents point to the relative success of Saudi Arabia and other key oil producers in adapting to the global credit crunch as evidence of the dawn of a new, more responsible era of governance in the Middle East.

Not everyone, however, sees the recent boom and bust as evidence of positive change in the region. In the eyes of many longtime observers the epic windfalls carry renewed risks for the future of the Middle East and for U.S. security, as high revenues exacerbate long-standing economic distortions and sociopolitical stagnation. Critics rightly note the role of petroleum rents in facilitating the region's democratic deficit and relentless resort to violence, a system in which, as scholar Fouad Ajami has written, "wealth comes to the rulers, they dispose of it, they distribute it to cronies, they punish and overwhelm their would-be challengers at home, and they use it to sustain adventures abroad way beyond the limits of their societies."[4] Rather than ushering the Middle East in the broader global economy, skeptics deride the region's recent spending spree as more white elephants designed to inflate investors' profits and rulers' vanity. Meanwhile, the central challenge facing the Middle East—harnessing the energy of its disproportionately young population—remains unmet as a result of an outmoded educational system, anemic job programs, and continuing reliance on expatriate labor. The collapse of the real estate market in Dubai and other overheated pockets of the regional economy has persuaded some skeptics that the region is destined to relearn the lessons of the 1980s.

Inevitably, the truth lies somewhere between these two hyperbolic scenarios and in fact incorporates elements of both. The oil boom has neither saved nor doomed the Middle East but rather opened up new possibilities and heightened existing problems in a way that makes the coming decade an especially critical one for the region and for the strategic environment for U.S. interests. In recent years, the region's petro-states have exhibited generally sound judgment in managing the boom—a notably positive development, given the pressures those states have had to bear in navigating their own internal contradictions and the changing international environment. However, alleviating and ultimately evading the traps associated with resource wealth will not come easily for the region in a period of mounting revenues.

The challenges facing the Middle Eastern governments in developing their societies and economies is not an abstract issue for Washington. The stability of the region is a fundamental U.S. security priority precisely because of its integral role in global energy markets. As a result, the United States—and more broadly, the international community—has a direct interest in ensuring that the region succeeds in navigating the inherent volatility of oil-based development. This chapter examines the implications, both positive and negative, of the long-term prospects for the Middle East's role in energy markets and concludes by offering policy recommendations for Washington to help ensure regional security and protecting U.S. interests in the free flow of energy.

The Context: A Future of High Oil Prices Driven by Demand

Over the past decade, as oil prices doubled and eventually raced past the seminal $100-per-barrel milepost, the issue of energy costs dominated

news headlines and political debates in the United States. Together with
the domestic U.S. credit crunch sparked by the subprime mortgage crisis,
escalating oil prices have been fingered as one of the main ingredients in a
vicious cycle that brought the U.S. economy and the world into a recession.
Considerable evidence suggests that factors beyond simple supply and
demand—such as the weak dollar, the frenzied activity of oil futures markets, ongoing instability in key suppliers such as Nigeria and Iraq, and
uncertainties about the possibility of new hostilities in Iran—played a role
in the price surge. Nonetheless, the underlying market conditions are also
directly relevant, both in understanding the current revenue stream accruing to producers as well as anticipating near- and medium-term trends.

Rising prices helped to facilitate massive development of additional
resources. A $50 billion investment over the past five years by Saudi Arabia
will bring the kingdom's production capacity from 9 million barrels per
day (bpd) to 12.5 million bpd, while smaller expansions are expected in
the United Arab Emirates, Nigeria, Qatar, Libya, and Angola as well as in
Russia, Azerbaijan, Kazakhstan, Brazil, and Canada. However, many of
the big prizes in terms of potential resource expansion remain off-limits for
the foreseeable future—largely a product of domestic politics and security
concerns in countries such as Iran, Iraq, Kuwait, Nigeria, and Venezuela.
In addition, the cost of developing new reservoirs is rising considerably as
a result of technology that provides access to previously unrecoverable
reserves and a price environment that drives production of formerly marginal supplies.

Ultimately, even with significant investment as well as increased conservation and innovation in alternative energy technology, the world is
facing increasing difficulty in keeping pace with the voracious global
demand for energy. As India, China, Gulf Cooperation Council (GCC)
countries, and East and South Asian countries continue to grow, albeit at
a slower pace because of the global recession, demand for oil has accelerated even as prices rise. Total worldwide demand grew by 10 million bpd
to 70 million bpd between 1977 and 1995—an eighteen-year interval.
By 2003, a mere eight years later, demand had reached 80 million bpd,
and while the global economic meltdown has seriously reduced expected
demand growth, current projections suggest that demand could reach
90 million bpd by the middle of the next decade. Two-thirds of future
increases in demand are expected to come from Asia, and most of that
demand will have to be satisfied by states that are members of the Organization of Petroleum Exporting Countries (OPEC), as supplies from pro-

ducing countries that are not members of OPEC are expected to plateau over the next five years.

To meet rising demand and cover natural declines in its own reservoirs, OPEC will have to produce at least 3 million bpd more each year, a challenge that one oil analyst describes as "an impossible task."[5] Rising prices and the global economic slowdown have mitigated those pressures, but given the exploding size of the Asian middle class and the short-term price inelasticity of demand for fuel, only a miraculous technological breakthrough or catastrophic changes in the Chinese and Indian economies will significantly alter the near-term demand picture. The burden—and the rewards—of the new global environment for energy will fall particularly to the Middle East's mature producers, thanks to the geological fluke that has made the Persian Gulf home to two-thirds of the world's proven oil reserves and approximately one-third of its natural gas. The International Energy Agency estimates that the primary sources for any new production sufficient to meet medium-term demand from a growing Asia will come from Saudi Arabia, Iran, and Iraq.[6]

The Opportunities: How High Oil Prices Are Stabilizing the Middle East

Although the consequences of escalating oil prices have been problematic for many countries around the world, the strategic position and natural resource inheritance of the Middle East has meant that it is uniquely positioned to benefit from price increases. Moreover, the region used the recent price spike to reduce debt, stockpile savings, and invest wisely—actions that it did not take during the 1973–85 oil boom—while intensifying long-needed structural reforms. Complemented by such wise management and external encouragement as needed, the broader ripple effects of the boom may generate some of the building blocks for a better future for the region.

The report card for the region's development in the aftermath of previous price spikes was notably dismal. One representative overview of the first two decades of epic oil wealth catalogues "the low returns to OPEC investments, useless white elephant projects, resource waste and moral corruption," adding that the "real pity" was that even after twenty years of accumulating massive resource wealth, most petro-states were "still not on a secure path toward sustained growth and prosperity."[7] At home, massive investment in physical and social infrastructure and economic development and diversification generated progress but overall relatively poor returns: educational systems that produced graduates unprepared for the job market, bloated and inefficient public sectors, lower average growth rates than during the pre-OPEC period, highly subsidized and unproductive non-oil sectors, and a declining share of world trade. Externally, the region directed most of its oil investment toward the United States, thanks in part to the unique openness of the U.S. economy, the shared security interests between most of the region's oil producers and Washington, and the assiduous U.S. government efforts to "recoup the American dollars flowing toward oil-producing capitals."[8]

In contrast, the early indications from the Middle East's experience during the recent price escalation as well as during the subsequent crash in prices offer some reasons for optimism that sounder judgment and more sober planning for optimizing the rewards of the region's resource wealth may yet prevail. Governments exhibited greater prudence by saving more and paying down debt; Saudi Arabia and Kuwait managed to reduce their external debt by more than half by 2005, from 97 percent to 41 percent of GDP and from 32 percent to 17 percent of GDP respectively.[9] During the first few years of the recent price escalation, the Saudis and other governments continued to base their government budgets on revenue expectations of $25 per barrel, only shifting upward well after that price band had become quaintly obsolete. Between 2002 and 2005, Middle East oil producers spent on average one-third of their windfall revenues, compared with 75 percent during previous oil booms. That forethought put them in a better position to weather the sudden and dramatic decline in prices that accompanied the global recession and helped to facilitate their quick and largely effective response to the downturn.

Massive budget and current account surpluses are providing unprecedented liquidity in financial markets, in turn boosting investment and growth. The oil boom has corresponded to an overall expansion of employment opportunities in the region and a decline in the regional unemployment rate from 15 percent in 2000 to 12.7 percent in 2005.[10] Many of the larger states appear to have placed a premium on projects with real potential to absorb the region's fast-growing labor market. Saudi Arabia's $200 billion investment in new "economic cities" is intended not just to compete with traditional hubs like Dubai and Bahrain but also to improve large-scale employment prospects, as the 2,000 planned factories and 800,000 planned jobs of King Abdullah Economic City and the 1.3 million projected jobs for an agribusiness city in Hail might suggest.[11] The GCC states have approximately $1 trillion in infrastructure investments already in the pipeline—and as much as double that amount if all the announced projects are actually launched.[12] Much of the investment in power generation, water desalination, education, and housing is desperately needed to support the rapidly growing population and to compensate for a legacy of domestic underinvestment compared with domestic investment by other middle-income economies such as Brazil, Russia, India, and China.[13]

The region is becoming far more globalized; foreign direct investment (FDI) in the Middle East expanded by more than 200 percent between 2001 and 2006—a tenfold increase in the region's proportion of global FDI.[14] Multinationals are racing to gain a foothold in this fast-growing region, such as Nasdaq's 2008 acquisition of a one-third stake in the Dubai International Financial Exchange. At the same time, Gulf states' foreign assets have more than doubled since 2003, with estimates ranging from $1.8 trillion to $2.4 trillion,[15] and their portfolios are much more diversified than ever before, with a greater focus on their own neighborhood as well as East and South Asia in spreading their largesse. These changes reflect a variety of factors: first and foremost, the evolving international market for capital means that in seeking to invest their windfalls, oil producers have far greater options today than they did during OPEC's earliest heyday. Moreover, turmoil in the region's relationship with Washington has slightly dampened Arab enthusiasm about U.S. investments, whereas markets closer to home offer the comfort factor of cultural and linguistic commonalities along with real opportunities for growth.

Whatever the rationale, the regional consequence is that the impact of the oil boom extends within the region far beyond the oil-rich states. Intra-Arab investment tripled between 2000 and 2005, and at least 11 percent of Gulf foreign investment since 2002 has remained within the region, particularly in North Africa.[16] "Gulf investors are going very big on North Africa," one hedge fund manager told a reporter in 2008.[17] The numbers may still be paltry relative to overall flow of revenues gushing into the Gulf, but for the recipients, the newfound regional interest can be decisive, particularly for economies still transitioning away from the heavy hand of state control. The United Arab Emirates invested $3 billion during 2007 in Egypt alone—a country whose stock market now draws 30 percent of its investors from the Gulf—and has made commitments to Morocco in the range of one-third the country's GDP. By 2020, overall Gulf investment in the Arab world could reach $750 billion—four times as much investment as occurred between 2002 and 2006.[18]

Examining the fast-rising Gulf interest in local financial products, where private regional investors now park at least one-quarter of their portfolios (as opposed to 15 percent in 2002), one consulting firm suggests that this investment is creating "a virtuous development cycle" that can strengthen local capital markets and help them mature.[19] One can reasonably extend that conclusion to the broader economic vitality of the region, since the multiplicity of alternatives available to the Gulf countries will minimize their tolerance for any business opportunities found in a poor investment climate. The growing interest in intraregional investment also holds out the prospect of not only facilitating further privatization and economic liberalization across the region but also mitigating some of the region's festering conflicts. Although the wealthy states of the region are rightly criticized for their stingy support of the Palestinians,[20] Gulf wealth played a significant role in rebuilding Lebanon after its civil war and in helping to stabilize its government and economy in the aftermath of the 2006 conflict with Israel and ensuing internal crises.

New relationships with Asia have reinforced the new eastern orientation to regional economic interests. That reflects a natural extension of the energy flows from the region; two-thirds of Gulf oil exports go to East or South Asia, which rely on the Gulf for at least that much of their oil supplies. Eleven percent of Gulf foreign investment since 2002 has been in Asia—a proportion that is expected to double by 2020.[21] Middle East investors (not including private equity firms) bought $20–30 billion in Asian assets in 2007, and trade between the two regions doubled between 2000 and 2005, with a tripling of exports from China, India, and Pakistan to the Persian Gulf. The changing vector of the region's economic interests raises a host of diverse issues and concerns for U.S. policy, but at a basic level this trend speaks to an unprecedented interdependence of nations and global integration that will enhance regional stability in the long term.

Beyond changing their spending patterns, regional governments are undertaking other real reforms, including the Saudi accession to the World Trade Organization, a new legal framework for corporate activities in the UAE, and a sustained effort to liberalize the Egyptian economy, which boosted foreign direct investment in Egypt from $300 million in 2003–04 to $6 billion in 2005–06. Even in Iran's disastrously mismanaged economy, high oil prices forced an unprecedented effort to address the long-standing distortion of gasoline prices produced by state subsidies through both rationing and substitution programs. A healthy competition for investment dollars and diversification has begun to transpire, pitting the pioneering

Dubai model against rivals in Abu Dhabi, Doha, Bahrain, and the rest of the region. There is a much keener focus on the overall climate for investment—what one Saudi official describes as the "soft infrastructure to help the business environment prosper"—that will help generate innovation and rising standards.[22] These trends have generated growth in economies that, at the end of the first major oil boom, actually contracted.

Ironically, the global economic crisis has provided a powerful test of the fitness of the regional economies and the soundness of most governments' strategies for managing the boom. While the vast oil price plunge—from $145 per barrel at its summer 2008 high to $33 per barrel only six months later—has eroded the glitzy growth rates for most of the Middle East, the impact has been considerably less than in other parts of the world. That reflects the relatively limited exposure of most regional banks to the credit crunch that overwhelmed so many American and European banks. Having paid down debt and engaged in genuine economic reforms, the region was well-positioned to ride out even a drop of that magnitude. In addition, a number of governments acted quickly to restore liquidity and shore up investor confidence in regional banks and stock markets. One of the most important vehicles for helping to address the adverse impacts of the global turmoil were the much-vilified sovereign wealth funds, which "played a significant stabilizing role domestically and abroad" by moving quickly to help shore up local bank shares and stock markets.[23]

The one exception, however, came in Dubai, whose heavy debt burden and reliance on real estate as a major driver of the local economy made it especially vulnerable to the ripple effects of the crisis. Still, even there, the economic turmoil may have a silver lining by bursting seemingly uncontrollable price escalation and forcing the cancellation or suspension of at least $75 billion in new projects. In the long term, the correction will impose new checks on corruption and speculation-driven growth and may encourage Dubai's brand-conscious leadership to adopt a more sustainable strategy for the long term. "We are going to tighten our belts, roll over and pay off debt, and be really trim over the next year," conceded the chairman of the emirate's splashiest property developer in January 2009.[24]

Developments outside the region have also positively shaped the context for social and political freedoms in the Gulf. As the Indian economy has liberalized and enjoyed record growth in recent years—itself an important factor in the current oil price equation—domestic demand for labor has intensified significantly. The new competition for human capital in India has helped generate higher standards and expectations from one of the most important labor sources for the Gulf. To some extent that has been expressed by implicit declines in labor supply; an Indian businessman noted in an interview posted on an Arab business website last year that "where you could have 10 laborers in the past, maybe you will only have five. Contractors will need to ensure their long-term sustainability by offering them a career path, good wages and living conditions."[25] The Indian government in 2007 set a minimum wage for overseas unskilled laborers, a move that was mirrored by the Philippines and Bangladesh.[26] Those moves are small but crucial steps in helping to improve the living standards and legal protections for the region's expatriate workers, its most vital as well as most vulnerable population.

Wealth is also generating much-needed investment in human capital. In addition, the Gulf states have launched massive new educational initiatives, building more than a dozen new campuses of U.S. universities and opening art galleries, media centers, and an array of cultural institutions at an investment of more than $20 billion a year.[27] The Gulf states have made it clear that they are prepared to import the very best of Western education and culture—from the Louvre to the Ivy League—and in most cases they have agreed to adhere to the rigorous standards and cultural norms of the home institutions themselves, such as by providing coed facilities and meeting U.S. accreditation requirements. The scope of such educational undertakings, coupled with a new focus on promoting entrepreneurship,[28] can help address one of the most important underlying risk factors for the regional environment, the demographic time bomb.

The Threat: How High Oil Prices Endanger Regional Energy Security

It would be tempting to view the latest avalanche of revenues and investment in the Middle East as an antidote to its manifold internal and external challenges. However, more than any other region of the world, the Middle East has long stood as a testament to the limitations of wealth in generating good governance and sustainable growth. As a result, amid the current boom times lies considerable reason to fear that the new global energy balance—in which demand is likely to sustain high prices for the near- to medium-term future—will only exacerbate the region's existing tendencies toward extremism, corruption, unrest, and intrastate violence. Under such a scenario, the perverse consequence of the new oil boom could be a Middle East that is far wealthier but even more unstable than it is today, with disturbing implications for the rest of the world's increasing reliance on Gulf oil and gas.

The reason for this prospective paradox is the well-documented link between resource wealth, growth, and autocracy, a function of the very mixed economic and political implications of resource wealth. Oil exploration and development is a highly capital-intensive industry that tends to create export enclaves without sufficient employment or related industrialization to promote balanced or sustainable development. States dependent on resource revenues are subject to intense fiscal volatility, wage and balance of payments distortions, and limited positive links in terms of economic and social development.[29] Paradoxically, given the perception of bounty, resource wealth is associated in reality with lower rates of economic growth and development.

Politically, a disproportionate reliance on external rents historically has distorted the political process by divorcing the state from any meaningful social accountability, reinforcing instruments of repression, spawning corruption, and eroding checks and balances. The state's primary role vis-à-vis society becomes a distributive one, and the result is a corrosion of formal institutions and the reinforcement of patronage.[30] A number of academic studies have demonstrated that oil-rich states, besides creating internal distortions, tend to be more likely to engage in conflict and spend more on security and maintaining larger armies than states that are not dependent on their oil resources.[31]

There are no simple solutions to the problematic consequences of resource revenues. Democracy is not among them, according to some scholars. Resource rents facilitate the typically preexisting patterns of patronage politics and erode the checks and balances, such as an open press, that might constrain traditional patterns of influence-seeking and revenue distribution. As a result, resource-rich governments fail to create the kind of public infrastructure that is beneficial to the development of competitive politics—or, for that matter, for economic growth. Scholars Paul Collier and Anke Hoeffler have demonstrated that "in those developing societies where the state has most command over resources, the democratic process has been least effective at controlling them for the public good."[32] As a result, introducing competitive elections in resource-rich societies has tended not to produce durable democracies or better management of the national wealth.

According to Stanford University political scientist Larry Diamond, none of the twenty-three countries that currently derive at least 60 percent of their export revenues from petroleum qualify as democracies and "all of the oil-rich countries of the world remained under or returned to authoritarian rule after 1974 and the third wave of democratization."[33] That includes all of the Middle East's major oil producers and extends more broadly across the region, where Freedom House's 2007 annual report designates only Israel as "free," with Bahrain, Lebanon, and Yemen categorized as partly free.[34] The region's proclivity for armed conflict is all too well established, from the epic warfare between Iraq, its neighbors, and several global coalitions, to the protracted failure of peacemaking between Palestinians and Israelis, to the persistence of terrorist violence against peoples and states from North Africa to Yemen.

Given this background, the forecasts of potential negative fallout from the region's renewed influx of revenues have obvious resonance. Recent years have brought more open elections and representative institutions to a number of Middle Eastern states, but considerable evidence suggests that those advancements have not fundamentally altered the authoritarian bargain that has long prevailed in the region, particularly in the oil-rich states.[35] The improving economic fortunes of the region will likely facilitate the perpetuation of that bargain, since, as Thomas Friedman has opined in the *New York Times,* "as the price of oil goes up, the pace of freedom goes down."[36] The logic appears to be borne out by the experience of countries such as Iran, where a dip in oil prices to as low as $10 per barrel during the late 1990s coincided with a president who championed a "dialogue of civilizations." In recent years, with oil prices careening to record highs, his successor spews anti-Israeli invective and oversees a new era of internal repression and international provocation.

Political reform carries its own substantial risks, but the relative dearth of meaningful steps toward greater accountability and popular participation creates significant uncertainties for the region's future—particularly in those states, such as Egypt and Saudi Arabia, that are poised to undergo rare changes in leadership in the near term. Managing those transitions may be rocky, and the surfeit of oil revenues may only complicate the process by facilitating corruption, entrenching privileged networks of power, and reducing incentives for good governance and rule of law, all of which would rebound negatively for economic development. Rather than the virtuous cycle described in the previous section, a future of enduring high oil revenues in the Middle East could generate the worst possible outcome for the region and for global interests in regional energy security: a region dominated by undemocratic and predatory regimes, sustained by oil windfalls but inherently precarious. Setting aside the particularist ideology of Iran's Islamic Republic, the ascendance of Mahmoud Ahmadinejad and his brand of radical populism and economic malpractice may be a harbinger of the region's future.

The prospects for such a scenario are reinforced by the demographic realities that complicate the region's internal challenges. Even with a new flood of cash, it is not clear that any of the regional states beyond the tiniest Persian Gulf emirates can sustain the social contract that has under-pinned the long-standing bargain between the region's rulers and the ruled. Two-thirds of the population of the Middle East is under the age of thirty, which represents a historic opportunity for growth in the context of the region's expanding economies; alternatively, this disproportionately young population could trigger what one expert described as "double jeopardy: the economic and social exclusion of youth drains growth and creates social strife."[37]

To marshal their young human resources successfully, regional states will have to embrace forward-leaning policies and programs to create 80 million new, productive jobs by 2020, nearly all in the private sector, as well as implement the sort of comprehensive educational expansion and reforms necessary to produce a trained and competitive work force. Today, youth unemployment and underemployment are rampant. Within the Gulf states alone, the challenge is to create 280,000 jobs per year to absorb new entrants to the labor markets—or 4 million new jobs by 2020 in a regional economy that currently employs only 4.8 million local citizens.[38] The boom has generated new private sector growth, but capacity remains grossly insufficient to meet the skyrocketing needs of most societies. Despite episodic political crises and the countervailing economic shocks of the oil price decline in the late 1990s and the current boom, the region's overall reliance on a primarily low-skilled, low-cost expatriate labor force has remained steady over the past decade at approximately 40 percent. Ambitious nationalization programs, including changes in the sponsor-ship system of some Gulf countries and a recent Saudi publicity campaign highlighting the labor minister's brief stint at a fast-food restaurant, have had only a limited impact. As a result, the impressive job creation targets remain largely aspirational, and the prerequisite structural changes—in particular, massive expansion and empowerment of the private sector—are still in their infancy. If these employment targets are not achieved, the specter of a burgeoning number of idle, frustrated youth looms on the horizon for the Gulf.

Compounding the political issues at stake in the region are real economic pressures that could exacerbate the task of maintaining stability at home. The scope of the windfall spending could overwhelm some economies, and population pressures and the associated infrastructure demands pose a little-discussed but very real set of hazards. The region already is suffering from widespread inflation—still modest by world standards but deeply worrisome within the context of local history—that has been sparked by massive spending, rising global food prices, and the local currency's peg to a weak dollar. In 2007 alone, construction costs in the Gulf rose by a shocking 30 percent.[39] Food prices are a major component of the challenge, and Egypt, Jordan, Yemen, and the UAE have each experienced food riots in recent years. Housing and real estate are the other dimension of the inflation problem: Dubai rents rose by 30 percent in 2006 and by another 17 percent in 2007.

Government efforts to address inflation by raising public sector salaries, building strategic food stocks, and enhancing domestic price subsidies have provided short-term relief for some beneficiaries but ultimately simply exacerbate the problem. Expatriate workers are especially vulnerable, as inflation stings both their own pocketbooks as well as the dollar-denominated remittances that they send back home. As a consequence, labor activism over pay and working conditions is on the rise around the region, and riots by expatriate laborers in Kuwait and the UAE have spooked the governments of both those states and their neighbors. The problems have only intensified as the worldwide recession has accelerated layoffs and repatriations of foreign workers.

The root of the inflation problem is twofold: most Gulf currencies are pegged to a steadily depreciating dollar, and the boom has sparked epic spending that tends to be highly dependent on imports. The dollar peg, which effectively forces the region's central banks to cut interests rates when they should be raising them, presents an especially knotty set of dilemmas. Any revaluation would cut both ways—alleviating domestic political and economic pressures while slashing the value of the GCC's trillions of dollars in offshore assets and causing friction with Washington.[40] Addressing the spending question is similarly fraught: deferring or scaling back the Gulf's approximately $1 trillion in infrastructure investments might mitigate the inflationary spiral but would leave the region's growing population even more vulnerable to power and water shortages. For all the justifiable criticism of the region's affinity for megaprojects, the efforts under way to modernize and expand the region's inadequate infrastructure represent a necessary response to a potential source of instability. Already, the challenge of meeting the youth bulge's basic needs—including power, water, health care, education, and transportation—is hampered by problematic shortages of both qualified contractors and concerns about credit availability and banking capacity.[41]

Beyond those pressures, other dimensions of the region's approach to capitalizing on the oil boom should give pause to expectations that the current price environment will inevitably facilitate a new era of peace and prosperity in the region. After decades of discussion, efforts to advance meaningful Gulf economic integration are still moving forward at a snail's pace, and as a result the region is missing real opportunities to optimize its resources, particularly in developing its much-needed power generation infrastructure. Gulf states' efforts to move their economies away from wholesale reliance on petroleum exports appear disturbingly interchangeable, raising concerns about overcapacity in aluminium, petrochemicals, and real estate ventures.[42] Also, the launch dates on several showpiece projects have been moved ahead precipitously—in the case of the King Abdullah Economic City in Saudi Arabia, a decade earlier than originally planned—raising concerns about excessive haste.[43]

Other initiatives, such as the massive education projects under way in the Gulf, may have adverse consequences that their enlightened backers never intended, by drawing much-needed talent and expertise away from the traditional centers of learning in the Arab world. "These are old societies with old roots," acknowledged Shafeeq Ghabra, who helped found the American University in Kuwait. "Even their cafes have been around for thousands of years. You can't replace that with shiny new classrooms and get the same level of depth."[44] In other words, the investments in educational infrastructure within the Gulf may only exacerbate and change the geographical vector of the region's long-standing brain drain, a problem that has already stripped the Arab world of one-quarter of its engineers and half of its doctors over the past thirty years.[45]

Skeptics point to the whiplash that buffeted the region in response to the global economic crisis as evidence of the inherently unsustainable and precarious development policies pursued by many Middle Eastern states. The wealthiest Gulf states had enough reserves to absorb the oil price crash, but governments whose margin for error was narrower did not fare as well—the slide below \$70 per barrel meant a return to government deficits in Iran and a deceleration of needed investments in Iraq. The impact has been even more severe for their resource-poor neighbors, which had benefited from the region's rising tides but now are the first and the hardest hit by financial tremors. Gulf investors have retrenched their positions in neighboring states, and remittances from expatriate workers have declined dramatically.

Inflation, budding bottlenecks, hyperdevelopment, lack of coordination, and unrealistic goals and timelines—all these maladies and distortions conjure comparisons to the overheated development schemes launched by Iran's monarchy in the years preceding the 1979 Islamic Revolution and by extension spark unease about the prospect of some future radical regime change in the region.[46] Moreover, as sustained high prices and increasing investment in alternative technology cut into demand for oil and gas, the region may yet again face the whiplash effect of declining government revenues but perpetually high popular expectations.

For obvious reasons, the region's internal political and economic challenges are directly relevant to the broader global economy and to U.S. security interests. The threats to infrastructure and transportation corridors are very real. Serious and sustained domestic unrest in any of the key Gulf oil and gas producers could disrupt export capabilities, as occurred in the aftermath of Iran's revolution. Even more ominously, internal destabilization could provoke terrorist attacks on oil export facilities and transportation routes; over the past several years, Saudi authorities have thwarted several such planned strikes by militants associated with al Qaeda. Epic oil revenues also generate renewed prospects for intrastate frictions, as empowered autocrats such as Iran's Ahmadinejad perceive themselves as invincible and the unlucky resource-constrained or indebted are left embittered and potentially emboldened in the manner of Saddam Hussein.

The Road Ahead

Ultimately, neither the judgments of the region's boosters nor those of its nay-sayers have it right. As the above overview suggests, the risks posed by the soaring price of oil to both the Middle East and U.S. interests are compelling and critical. However, the results need not be foreordained. The searing experience of prior oil price crashes in 1985–86 and 1997–98—when OPEC revenues dropped by more than three-quarters and by one-third respectively[47]—coupled with the unavoidable pressure of a young, globalized, and demanding population has in fact generated a determination among regional leaders to avoid another lost opportunity. "We want to learn from the mistakes of the '70s," one Saudi official acknowledged.[48] With the hindsight and the judiciousness demonstrated in the early stewardship of the current boom, there is some reason to believe that the past need not replicate itself.

Still, it is dangerous and irresponsible to rely on regional leaders' hard-earned prudence, their willingness to engage in modest top-down reforms, and the buoying effect of economic growth to ensure a durable pathway to a more secure and prosperous future for the Middle East. Contemporary regional history and politics offers little evidence that prosperity alone begets peace and stability or even that economic reform alone can generate political systems and cultures that are conducive to long-term stability.[49] Iran's cataclysmic revolution unfolded in the context of rapid economic growth made possible by an unprecedented bonanza of oil revenues, as well as corruption and pressures on powerful elements among the merchant community, while more recent developments in Tunisia and Egypt underscore the capacity of authoritarian states to liberalize their economies without yielding an inch of their tight political control.

The obvious conclusion to be drawn from this mixed forecast is that the region and the world must work cooperatively to maximize the opportunities presented by the oil boom by ensuring that the windfalls pay broader dividends. It will not be an easy task. Most of the rest of the world must find ways to cope with the short-term economic pain caused by epic oil prices, while doing much more to shift the energy balance through conservation and development of alternative energy sources and technologies. Focusing on the dilemmas of the boom's apparent "winners" understandably falls lower on the list of priorities. In addition, the challenge is complicated by the extent to which the changing environment for energy reduces U.S. leverage. For Washington, the oil windfalls—and the likelihood of their indefinite perpetuation—legitimately intensify the imperative of assisting the Middle East in navigating a path toward sustainable prosperity and meaningful political reform. Yet the U.S. ability to pressure and persuade is inevitably constrained by the dependence of both the U.S. economy and that of the rest of the world on the very commodity that is responsible for this epic regional bounty.

These contradictions were on full display during President George W. Bush's final tour of the region in May 2008. In Egypt, before an array of Arab political and business leaders, Bush issued a stirring appeal for reform and democracy, declaring that "too often in the Middle East, politics has consisted of one leader in power and the opposition in jail. The time has come for nations across the Middle East to abandon these practices, and treat their people with dignity and the respect they deserve."[50] The strength of the president's rhetoric was powerfully undercut by an entreaty that he had issued more quietly during an earlier stop on his trip: a request to Saudi leaders to expand oil production in order to lower the spiraling price of gasoline. That neither presidential request was likely to succeed only underscores the faltering influence of the United States and the depth of the challenge that the country faces.

For that reason, the Bush administration's successors will have to move beyond the soaring oratory about democracy and the feel-good programming that accompanied the much-vaunted "Freedom Agenda" to work with regional and individual leaders on the specific opportunities and threats that stem from the hyperwealth of the current oil boom. One potentially useful step entails reviving and upgrading a formal channel for dialogue and cooperation between Washington and the region on economic issues, such as the U.S.-GCC Economic Dialogue, which devolved to the U.S. Commerce Department and was abandoned in 2001. The particular challenges of the oil boom warrant the involvement of senior officials on both sides, along the lines of the strategic dialogue with China that has been led by the deputy secretary of state and the secretary of the treasury. That level of participation ensures that the channel transcends the standard focus on trade promotion, highlights the urgency of the shared interest of the United States and the Gulf states in seeing the region steward its oil revenues wisely, and helps ensure real buy-in from the relevant agencies on both sides. The dialogue should incorporate working groups composed of representatives of each side, tasked to address specific priority issues on an ongoing basis. The resumption of dialogue in some higher-profile format, along the lines of recent U.S.-China diplomacy on economic issues, would also assuage some of the umbrage within the Gulf—particularly in Riyadh—about the Bush administration's appropriate and astute decision to pursue bilateral rather than multilateral free trade agreements with the Gulf states.

Several important economic issues appear ripe for greater engagement among the Middle East, Washington, and other weighty international actors, including China, Japan, the European Union, and Russia. One is mutual concern about the increasing prominence of Gulf-based sovereign wealth funds in investing abroad; both the European move toward voluntary codes and the principles developed in recent discussions between U.S. Treasury officials and Gulf leaders suggest the utility of a broader effort to develop standards and mutual understanding as a means of avoiding politicization of this issue by either side.

A second topic worthy of focus is the direction and composition of regional foreign assistance programs. Historically, Washington has been prone to the temptations of "tin-cupping"—turning to oil-rich states on a case-by-case basis to fund development and reconstruction projects that have either political (Iraq, Afghanistan) or humanitarian (disaster relief) priority for the U.S. administration. This is an inefficient and unstructured approach that serves no party's ultimate interests, and the likely continuation of the large revenue streams to the Gulf merit a more systematic effort to identify areas of priority and potential cooperation.

Third, the United States, together with the international financial institutions, should initiate a much more intensive effort to facilitate greater regional economic integration. Short-term political obstacles have hindered long-standing interests in developing a regional power grid for the Gulf countries, which could lead to future electricity shortages in some countries as well as considerable unnecessary financial costs. A U.S. diplomatic effort similar to the dialogue suggested above and, as necessary, incentives should be deployed to ensure long-term interest in projects like this and those that would create similar linkages for water and transportation within the Gulf and across the Middle East more broadly.

Finally, Washington and other major oil-consuming nations should elevate the dialogue with the major oil producers in the Middle East about transparency in the petroleum sector. Combined with efforts to strengthen the capacity of indigenous institutions such as the media and parliaments, access to information about oil and gas revenues and spending is the most low-cost, high-return tool available for promoting good governance and accountability. A voluntary code known as the Extractive Industries Transparency Initiative (EITI) has generated widespread interest and buy-in from NGOs, governments, and companies but has failed to penetrate the Middle East, with the recent exception of Iraq. Given the political sensitivities around oil as a national patrimony within the region, the tendency toward secrecy is understandable, but ultimately it is detrimental to both the market and the political evolution of the region. No external actor can force the region to embrace transparency, but the United States can do more to impress upon the Middle East the valuable role of EITI and other efforts to enhance national accountability.

Of course, summitry is not equivalent to actual solutions, and the real objective of both regional leaders and the new U.S. administration is the promotion of good governance and enlightened management of the oil bounties accruing to the Middle East.

That will entail a comprehensive transformation of long-standing patterns of behavior by both public and private actors in the region. Washington can and must help the region navigate between the perils and possibilities of the new era for energy and oil revenues, but ultimately the region and its people must determine its fate.

Notes

1. Mohamed Fadhel, "Expat Workers Strike as Earnings Are Hit," *Gulf Times*, February 18, 2008, p. 9; "Car License Plate Sets World Record," *Gulf Times* (February 18, 2008), p. 9.

2. Moin Siddiqi, "Gulf Poised for Heady Surge into 2008 and Beyond," *Middle East* 387 (March 2008), p. 42.

3. Ibid.

4. Fouad Ajami, "The Powers of Petrocracy," *U.S. News and World Report*, December 19, 2007 (www.usnews.com/articles/opinion/fajami/2007/12/19/the-powers-of-petrocracy.html).

5. Marilyn Radler, "MEGC: Oil Price Strength Here to Stay; Demand Fails to Waver," *Oil and Gas Journal* 106, no. 14 (April 14, 2008), p. 28.

6. Nick Snow, "U.S. Foreign Policy Must Consider Changing Energy World," *Oil and Gas Journal* 105, no. 3 (January 15, 2007), pp. 39–40.

7. Jahangir Amuzegar, *Managing the Oil Wealth: OPEC's Windfalls and Pitfalls* (London: I.B. Tauris, 2001), p. 222.

8. Rachel Bronson, *Thicker than Oil: America's Uneasy Partnership with Saudi Arabia* (Oxford University Press, 2006), p. 124.

9. Pamela Ann Smith, "GCC Foreign Wealth Rises to $2 Trillion," *Middle East* 388 (April 2008), p. 37.

10. Paul Dyer, "Will the Oil Boom Solve the Middle East Employment Crisis?" Dubai School of Government, Policy Brief 1, November 2006, p. 2.

11. "How to Spend It," *The Economist*, April 24, 2008.

12. Smith, "GCC Foreign Wealth Rises to $2 Trillion," p. 37; "How to Spend It," *The Economist*.

13. Smith, "GCC Foreign Wealth Rises to $2 Trillion," p. 37.

14. Siddiqi, "Gulf Poised for Heady Surge," p. 42.

15. "How to Spend It," *The Economist*; Kito de Boer and others, "The Coming Oil Windfall in the Gulf," McKinsey Global Institute, January 2008, p. 6 (www.mckinsey.com/mgi/reports/pdfs/the_coming_oil_windfall/Coming_Oil_Windfall_in_the_Gulf.pdf).

16. de Boer and others, "The Coming Oil Windfall," p. 18.

17. Smith, "GCC Foreign Wealth Rises to $2 Trillion," p. 37.

18. Ibid.

19. de Boer and others, "The Coming Oil Windfall," p. 13.

20. Glen Kessler, "Arab Aid to the Palestinians Often Doesn't Fulfill Pledges," *Washington Post*, July 27, 2008, p. A16.

21. de Boer and others, "The Coming Oil Windfall," p. 18.

22. Faiza Saleh Ambah, "Saudis Look beyond Oil to New Economy in Desert," *Washington Post*, July 17, 2008, p. A1.

23. International Monetary Fund, "Regional Economic Outlook: Middle East and Central Asia," May 2009, pp. 10–11.

24. Simeon Kerr, "Emirate on the Ebb," *Financial Times*, January 30, 2009, p. 9.

25. Peter Sorel-Cameron, "Is the Gulf Headed for a Labor Crisis?" CNN Marketplace Middle East, March 14, 2008 (http://edition.cnn.com/2008/BUSINESS/03/14/labor.mme/index.html).

26. "Labour Pains in the Middle East," *The Economist*, March 31, 2008.

27. Zvika Krieger, "Desert Bloom," *Chronicle of Higher Education*, March 28, 2008, p. B7.

28. Stefan Theil, "An Arab Opening," *Newsweek International*, August 20, 2007.

29. Hazem Beblawi and Giacomo Luciani, *The Rentier State* (London: Croom Helm, 1987), explores rentierism in various states across the Middle East.

30. Samuel R. Schubert, "Revisiting the Oil Curse," *Development* 49, no. 3 (September 2006), pp. 64–70.

31. Paul Collier and Anke Hoeffler, "Resource Rents, Governance, and Conflict," *Journal of Conflict Resolution* 49, no. 4 (2005), pp. 625–33; Michael L. Ross, "Blood Barrels: Why Oil Wealth Fuels Conflict," *Foreign Affairs* (May–June 2008).

32. Paul Collier and Anke Hoeffler, "Democracy and Resource Rents," Global Poverty Research Group Working Paper, April 26, 2005, p. 18 (http://users.ox.ac.uk/~econpco/research/pdfs/Democracy-resource-rents.pdf). See also Paul Collier and Anke Hoeffler, "Testing the Neocon Agenda: Democracy in Resource-Rich Societies," Working paper, Department of Economics, University of Oxford, November 2007, p. 32 (http://users.ox.ac.uk/~econpco/research/pdfs/TestingTheNeoconAgenda.pdf).

33. Larry Diamond, *The Spirit of Democracy: The Struggle to Build Free Societies throughout the World* (New York: Times Books, 2008). Diamond's observation on oil as an inhibitor of democracy is borne out by statistical analyses, including Michael L. Ross, "Does Oil Hinder Democracy?" *World Politics* 53 (April 2001), pp. 325–61.

34. Freedom House, *Freedom in the World Country Ratings: 1972–2006* (www.freedomhouse.org/uploads/fiw/FIWAllScores.xls).

35. Steven Heydemann, "Upgrading Authoritarianism in the Arab World," Saban Center for Middle East Policy, Brookings Institution, Analysis Paper 13, October 2007 (www.brookings.edu/~/media/Files/rc/papers/2007/10arabworld/10arabworld.pdf).

36. Thomas L. Friedman, "The Democratic Recession," *New York Times*, May 7, 2008 (www.nytimes.com/2008/05/07/opinion/07friedman.html?scp=1&sq=friedman%20democratic%20recession&st=cse). See also "The First Law of Petropolitics," *Foreign Policy* (April–May 2006) (www.foreignpolicy.com/story/cms.php?story_id=3426).

37. Navtej Dhillon, "Middle East Youth Bulge: Challenge or Opportunity?" presentation to congressional staff, May 22, 2008 (www.brookings.edu/speeches/2008/0522_middle_east_youth_dhillon.aspx).

38. de Boer and others, "The Coming Oil Windfall," p. 9.

39. "Labour Pains in the Middle East," *The Economist*.

40. Siddiqi, "Gulf Poised for Heady Surge," p. 42.

41. Jeff Black, "Different This Time?" *Middle East* 388 (April 2008), p. 35.

42. "How to Spend It," *The Economist*.

43. Ibid.

44. Krieger, "Desert Bloom," p. B7.

45. Ibid.

46. Kenneth M. Pollack, "Drowning in Riches," *New York Times*, July 13, 2008.

47. Anthony H. Cordesman, *Energy Developments in the Middle East* (Westport: Praeger Publishers, 2004), pp. 122–26.

48. Ambah, "Saudis Look beyond Oil," p. A1.

49. For one excellent overview of the interconnections between economic and political reform in the Middle East, see Eva Bellin, "The Political-Economic Conundrum: The Affinity of Political and Economic Reform in the Middle East and North Africa," Carnegie Endowment for International Peace, Working Paper 53, November 2004.

50. President Bush's speech before the World Economic Forum, Sharm El Sheikh, Egypt, May 18, 2008 (www.whitehouse.gov/news/releases/2008/05/20080518-6.html).

CHAPTER THREE

How Much Does the United States Spend Protecting Persian Gulf Oil?

MICHAEL O'HANLON

How much does the United States spend on its military to defend the Persian Gulf region and, more specifically, to ensure the stable and orderly production of oil in and the flow of oil out of that region? Since the articulation of the Carter Doctrine in the 1970s, protection of the Persian Gulf has been a formal element of U.S. defense strategy. Even more vividly, in recent decades the United States has fought two major wars in and around Iraq and has maintained continuous military vigilance toward Iran.

The cost question is central in comparing the costs and benefits of different energy strategies. Even though there are other reasons for the United States to worry about security in the Gulf and broader Middle East region, beginning with the well-being of Israel as well as that of other friendly countries such as Lebanon and Jordan, it is largely the need for oil that drives the strong U.S. commitment to this theater.[1] In fact, many U.S. foreign policy interests argue for minimizing the U.S. military presence in the region to avoid stoking anti-Americanism and providing fodder for the likes of Osama bin Laden to allege that the infidel West has secret aims of controlling the Middle East.

The question of cost is difficult to answer, however. The main reason is that, apart from times like the present when a large fraction of U.S. combat forces are deployed within the broader Middle East (and funded largely by supplemental appropriations that can be separately identified

59

and analyzed), no major package of U.S. military units is dedicated to the Persian Gulf region alone. Central Command has specific command headquarters and basing arrangements in the Gulf, to be sure, but the former number only in the thousands of troops and the latter in the low tens of thousands (including naval forces in the Gulf itself). Together these quasi-permanent U.S. military contributions to the Gulf region cost 1 to 2 percent of the annual defense budget, or $5 billion to $10 billion a year—obviously far less than the actual amount spent on forces that might be and often are deployed to the Gulf, even if they are hypothetically usable for other regions of the world as well.

In fact, my best estimate is that the United States spends about $50 billion a year on the region's security *without* counting the costs of specific operations like the one under way in Iraq, which at present costs more than another $100 billion a year. Since the United States imports some 1 billion barrels of oil a year from the Persian Gulf, a simple calculation suggests an implicit subsidy of about $50 a barrel. If one assumes that about half of the subsidy goes toward gasoline, that amounts to about 50 cents a gallon of imported fuel—roughly comparable to other authors' estimates of the implicit subsidy. Arguably, though, the cost should be distributed across all gasoline, not just imported gasoline, since what is being implicitly subsidized is an overall system of oil use. In that case, the cost per gallon would be closer to a dime.[2] Reaching such an estimate requires numerous judgments and simplifying assumptions to be made that require explanation and rationale. This chapter attempts to provide them, beginning with a short explanation of the various budget categories used by the Department of Defense (DoD) to categorize and subcategorize its overall expenditures.

Basic DoD Budget Categories

The U.S. military breaks down its official overall budget in several ways. Two basic methods show spending by title and by service. Another method, devised by former secretary of defense Robert McNamara, subdivides spending by what he called military "programs." Rather than allocate the defense budget on the basis of branch of military service or type of activity, he sought to use broad functional categories, including strategic nuclear capabilities, main combat forces, transportation assets, administrative and related support activities, National Guard and reserve forces, intelligence, and several smaller areas of expenditure. This method is itself

TABLE 3-1. DoD 2009 Budget Authority Request by Title
Constant 2008 dollars, in billions

Military personnel	125.2
Operations and maintenance	179.8
Procurement	104.2
Research, development, testing, and evaluation	79.6
Military construction and family housing	24.4
Management funds, transfers, and receipts	2.2
Total	515.4

Sources: Under Secretary of Defense (Comptroller), *Military Personnel Programs (M-1), Operation and Maintenance Programs (O-1), Department of Defense Budget, Fiscal Year 2009* (Department of Defense, February 2008), pp. 18, 20; Under Secretary of Defense (Comptroller), *Construction Programs (C-1), Department of Defense Budget, Fiscal Year 2009* (Department of Defense, February 2008), p. iv; Under Secretary of Defense (Comptroller), *Procurement Programs (P-1), Department of Defense Budget, Fiscal Year 2009* (Department of Defense, February 2008), p. II; and Under Secretary of Defense (Comptroller), *RDT&E Programs (R-1), Department of Defense Budget, Fiscal Year 2009* (Department of Defense, February 2008), p. II.

not perfectly revealing or accurate. For example, many military forces can be used for both nuclear and conventional operations; should they be viewed then as strategic nuclear capabilities or main combat forces? Another thorny analytical problem is how to allocate expenditures for equipment first bought for active forces but later transferred to the reserves. Moreover, these categories are sufficiently broad that, even if accurate, they may have only a modest bearing on critical policy choices. Nevertheless, they do provide at least an order-of-magnitude sense of how different types of military objectives or mainstream activities translate into costs.

In tables 3-1, 3-2, and 3-3, all costs, which are given in billions of constant 2008 dollars of budget authority, reflect the Bush administration's request for fiscal years 2008 and 2009 (running from October 1, 2007, through September 30, 2008, and not counting any war costs).[3]

Available documentation, updated each February with the Pentagon's budget request, provides a great deal of detail for analysts trying to dissect the military budget.[4] For example, within the military personnel accounts, information on travel and moving allowances can be found along with information on officer pay versus enlisted pay, current salaries versus future retirement benefits of current troops, and active troop compensation versus reserve troop compensation, to name but a few subcategories. Within the procurement budgets, in addition to breakdowns by service, there are subcategories for aircraft, vehicles, ammunition, missiles, and other asset groupings.

T A B L E 3 - 2 . DoD 2009 Budget Authority Request by Service
Constant 2008 dollars, in billions

Army	140.7
Navy	149.3
Air Force	143.9
DoD-wide	81.6
Total	515.4

Source: Under Secretary of Defense (Comptroller) Tina W. Jonas, "Fiscal Year 2009 Budget Request: Summary Justification," Department of Defense, February 4, 2008, p. 8.

Which category is most useful for understanding a given policy challenge or framing a given policy choice depends on the issue at hand. Familiarity with the above breakdowns can answer the occasional policy question. For example, imagine that someone wishes to know whether the country should move to a smaller but more mobile force posture. One way to find out might be to double the budget for U.S. mobility forces, using

T A B L E 3 - 3 . DoD 2008 Budget by Program[a]
Constant 2008 dollars, in billions

Strategic forces	10.4
General purpose forces	189.1
Command, control, communications, intelligence, and space	72.1
Mobility forces	13.4
Guard and reserve forces	36.0
Research and development	49.9
Central supply and maintenance	21.5
Training, medical, and other	63.5
Administration	14.2
Support of other nations	2.1
Special operations forces	9.2
Total	481.6

Sources: See Under Secretary of Defense (Comptroller), *National Defense Budget Estimates for FY 2008*, pp. 1–2, 81; Office of Management and Budget, *Budget of the U.S. Government, Fiscal Year 2008: Historical Tables* (Government Printing Office, 2007), pp. 89, 164; and Allen Schick, *The Federal Budget: Politics, Policy, Process* (Brookings, 2007), p. 57.

a. Here the figures add up to a slightly different total because what is presented is total obligational authority, not budget authority. The difference in these two concepts is quite small and unimportant for our purposes. Another detail worth noting here concerns the distinction between discretionary budgets and overall, total budgets. Discretionary funds have to be appropriated each year by Congress. Overall budgets also include mandatory programs and spending, which do not require annual attention (entitlements are the largest example of mandatory programs in the federal budget). Almost all military spending is discretionary. And mandatory accounts can be positive or negative as they can involve trust funds, user-fee programs, and the like. For example, in 2008 the administration's request for all DoD funding was $643.7 billion; the discretionary request was for $647.2 billion, meaning that the mandatory funding request was "negative."

savings from a smaller combat force structure to fund expanded transportation programs. Someone wondering whether deep cuts in nuclear programs could save a great deal of money could get some information from the above. (However, it is important to remember that Department of Energy nuclear warhead costs and many missile defense costs are not captured in the strategic forces category used above for such purposes.)[5] If the question is how a 5 percent across-the-board increase in military compensation would affect the defense budget, the above information on title and category may come in handy. (Incidentally, civilian pay for DoD employees, which totals a bit more than $60 billion a year at present, is located within the budget for operations and maintenance).[6]

Furthermore, if one is trying to evaluate the hypothesis that Pentagon politics makes it hard for the relative budgets of the Army, Air Force, and Navy to change very much (because each service opposes cuts to its share of the budget), mapping trends in the budget shares for each service can help evaluate that hypothesis. The hypothesis is probably more right than wrong, by the way—but that does not mean that service shares should be changed recklessly just to overcome the Pentagon's inertia. For example, at first the Rumsfeld Pentagon wanted to cut back substantially on the Army and use the savings to invest in defense transformation and technology. In light of the Iraq experience, however, it seems clear now that doing so would have been a major mistake.

Generally, more refined budgetary tools are needed for these and other purposes. The above budgetary frameworks are informative, but they are not analytically powerful; because of that, I go further in subsequent discussions. First, however, the Iraq war's costs and supplemental appropriations are analyzed.

The Wartime Supplementals

For 2008 the Pentagon requested a total of $189 billion in supplemental costs for wartime operations. Unusually, it placed the first $142 billion of that total directly into the main defense budget request. The United States usually does not have funds for major operations in its normal defense budgets, primarily because Congress guards its prerogative to fund actual operations carefully and jealously, denying any upfront funding for major operations and requiring the executive branch to come back to it for supplemental funding should the nation go to war. In 2008, Congress suggested combining the budgets for the ongoing wars in Iraq and Afghanistan within the normal budget, perhaps because the wars had gone on for so long or perhaps to force the president to acknowledge in a transparent fashion how expensive his defense policy had become for the nation.[7]

To determine how supplemental costs can be broken down, it is analytically useful to examine the $142 billion initial request for overseas missions in 2008 in more detail. Exactly half, or $71 billion, was allocated for operations while another $38 billion was for repair, replenishment, and general reconstitution of the force as it cycled back from Iraq. Another $11 billion was for force protection, including armored equipment, and $4 billion more was for other activities (beyond armoring vehicles), specifically countering improvised explosive devices. Smaller amounts of funding were devoted to the Iraqi and Afghan security forces, military construction activities, and classified accounts. The additional $47 billion requested later in the year included $6 billion more for operations; almost $20 billion for force protection, including large numbers of mine resistant, ambush-protected (MRAP) vehicles; $9 billion for reconstitution of equipment; and several billion more for U.S. training of Iraqi security forces and for U.S. military facilities in the region.[8]

Under another methodology, the initial 2008 operations budget of $142 billion allocated $110 billion to Iraq, $26 billion to Afghanistan, and $6 billion to miscellaneous as well as classified purposes. If one adopts Pentagon jargon utilizing "military title" categories (described further below), the $142 billion included about $17 billion for military personnel costs, $72 billion for operations and maintenance, and $33 billion for procurement, with smaller amounts for other activities.[9]

Through August 2007, Congress had appropriated $610 billion for the "broadly defined" global war on terror, including $538 billion for Department of Defense operations, $30 billion to fund Iraqi and Afghan security forces, and $42 billion for other departments and agencies. About $450 billion was used for Iraq, $127 billion for Afghanistan, and the remaining $32 billion largely for Operation Noble Eagle, to help protect the homeland. With the subsequent $189 billion in combined funding for 2008, the total since 2001 reached $800 billion. (In comparison, in 2008 dollars, the Korean War cost about $470 billion, Vietnam about $665 billion, and Desert Storm about $90 billion, with 90 percent of the latter costs paid by U.S. allies). In rough terms, total funding for Iraq reached about $600 billion; for Afghanistan, $150 billion; and for DoD homeland security efforts, $50 billion.[10]

By 2007, the marginal costs of Afghanistan and Iraq combined had reached $10 billion a month—$8.6 billion for Iraq, $1.4 billion for Afghanistan. By marginal costs, I mean costs above and beyond those already incorporated for the forces in question in the standard defense budget (such as regular salaries and regular training costs and health care needs). Already by 2006, the marginal cost per deployed troop averaged over $500,000 a year. This was more than twice what had been projected for the war back in 2002. Even once costs for activities such as training Iraqi security forces are removed, costs per U.S. troop deployed have now exceeded $400,000 a year. DoD has been notorious for a failure to understand deployment costs accurately in the recent past; for example, in the Bosnia mission, initial estimates for the cost of deploying 20,000 troops to the region for a year were $1.5 billion to $2 billion, but actual costs were at least twice the upper bound.[11]

Why are the numbers for the current wars in Afghanistan and Iraq so much higher, on a per-troop basis, than those for past wars or for earlier estimates for these wars? It is one thing to have some uncertainty in projections of war costs because it is unclear how long a war will last or how hard the fighting will be. For example, before Operation Desert Storm in 1991, the Congressional Budget Office estimated that costs could run between $43 billion and $135 billion (translated into 2008 dollars).[12] It is something else to be off by 100 percent or more when the duration and troop strength of a given mission are not in doubt. The primary explanation is that most costs besides those for troop benefits (such as hostile fire pay) have escalated far beyond what was predicted. Military facilities have been developed to be high-tech, comfortable, and useful over a sustained period. Equipment has been worn down by intense operations and damaged by enemy action far more than forecast. Contractors have been hired to support operations in very large numbers. Finally and quite unabashedly, DoD has added a number of costs for activities not strictly related to the war—such as restructuring its Army brigades—in supplemental requests since 2005.[13]

The easiest way to see this is in the funding history of wartime activities. Supplemental procurement funding, which averaged only about $10 billion annually through 2005, rose to $25 billion in 2006 and to $51 billion in 2007—and the funding request in 2008 was for a whopping $72 billion. Supplemental operation and maintenance costs also have skyrocketed, doubling since 2004 from about $45 billion to almost

$90 billion a year. Among other things, many of the Army "reset" costs to return equipment to previous, pre-war condition were funded out of these accounts, and the scale of the operations in Iraq and Afghanistan in 2007 and 2008 grew somewhat, but relatively modestly. (In contrast, supplemental military personnel costs have held relatively steady since 2003, averaging between $16 billion and $18 billion a year.)[14]

Kaufmann's Geographic Breakdown of U.S. Defense Costs

While it is important to have a grounding in the Pentagon's basic budget categories, the categories summarized above do not suffice when one seeks to analyze policy alternatives. The breakdowns by service and by title fail to describe very much the missions to which defense dollars are devoted. Even the McNamara program elements fail to give any insight into the force structures designed to carry out those missions or the per-unit costs associated with them.

In recognition of this dilemma, longstanding Pentagon adviser and Brookings scholar William Kaufmann created two additional breakdowns of force structure, weapons purchases, and related Pentagon costs that can be used to complement the McNamara method and often provide more useful analytical tools. One approach, subdividing costs by the world's geographic regions, is employed in the following analysis.

Kaufmann's geographic approach views U.S. military missions as focused primarily on various overseas theaters—Europe, the Atlantic sea lanes, the Far East, the Persian Gulf, Latin America, and Africa. Most combat formations are assigned accordingly, though some are attributed to U.S. territorial defense or to missions such as nuclear deterrence and intelligence. Kaufmann's taxonomy is similar to McNamara's, including about ten main categories. As with McNamara's, his allocations are constructed so that the total equals the aggregate defense budget.

The basic logic of Kaufmann's allocation scheme is simple and appealing and, if accurate, provides a clear method of assessing the fiscal implications of various U.S. security commitments, such as protection of Persian Gulf oil or of key allies such as Japan, Korea, and the countries of western Europe. The critical analytical question, however, is whether it is right.

Indeed, while not lacking merit, Kaufmann's geographically oriented defense budget breakdown is probably the most controversial of the major methods considered here. Certain military assets are designed primarily for one type of operation in just one or two places—for example, frigates designed to protect ships as they traversed the Atlantic and Pacific during the cold war and aircraft carriers and Marine expeditionary units routinely deployed to specific regions such as the Persian Gulf and Western Pacific. In such cases, it is not difficult to apportion costs on a regional basis, at least roughly. For example, examining their homeports and typical deployment patterns can help elucidate whether ships and ship-based Marines should be assigned an Atlantic/Mediterranean or a Pacific/Indian Ocean designation. During the cold war, because Europe was the primary heavy combat theater for air-ground operations, it was logical to attribute the costs of most Army and tactical Air Force units to that region. After the cold war, the focus for such capabilities shifted to the Persian Gulf and Korean peninsula, as Pentagon documents that guided overall Pentagon strategy and resource allocation, such as the Clinton administration's 1993 "Bottom-Up Review," explicitly reveal. At least to some extent, therefore, the above examples show the efficacy of using Kaufmann's regional methodology.

Kaufmann's last breakdown was done in 1992, when the Pentagon worried much more than it does today about a possible Russian resurgence and the resulting hypothetical danger to countries like the Baltic states. Consequently, Kaufmann estimated that a substantial fraction of the overall defense budget was for the defense of Europe. It is not clear whether he would reach the same conclusion today. Kaufmann's calculations are shown in table 3-4, displayed as percents of the overall Department of Defense budget. Only the budgets for nuclear forces and for national intelligence and communications are not divvied up by region. Moreover, the share of the defense budget focused on the Western Pacific—with an eye not only toward North Korea but also rising Chinese power and the general ascendance of Asia in economic and strategic terms—conceivably might change if Kaufmann were to redo his estimates today. In addition, the United States has been more active in Africa over the past fifteen years, beginning with the ill-fated Somalia mission but also including refugee relief in Central Africa, counterterrorism cooperation with countries in the Sahel, and now the creation of the new Africa Command (Africom). Nonetheless, Kaufmann's numbers reflect his initial assumptions about the state of the post–cold war and post-Soviet world and therefore still have some relevance today.

The resulting budget tools that Kaufmann created are most useful in trying to assess how much the country spends defending particular U.S. interests overseas and specific interests of our allies—and how much it

T A B L E   3 - 4 . Kaufmann's Estimates of DoD Spending by Geographic Region under the "Base Force" of 1992

Percent

Strategic nuclear deterrence	15
Tactical nuclear deterrence	1
National intelligence and communications	6
Northern Norway/Europe	5
Central Europe	29
Mediterranean	2
Atlantic sea lanes	7
Pacific sea lanes	5
Middle East and Persian Gulf	20
South Korea	6
Panama and Caribbean	1
United States	3

Source: William W. Kaufmann, *Assessing the Base Force: How Much Is Too Much?* (Brookings, 1992), p. 3.

might save if it reduced certain commitments (or how much more it might spend if it increased commitments). This framework can help shape debates over allied military burden sharing, in discussions of how much the United States spends defending Persian Gulf oil, and so forth.

However, it would be a mistake to take Kaufmann's framework too literally. The U.S. armed forces do not in fact create force structures devoted to just one region. It is very rare to have a combat formation that can be used in only one part of the world. To be sure, some head-quarters capabilities, some planning staffs, and some intelligence assets are devoted to a specific region; furthermore, the occasional combat unit, such as the Army's 2nd infantry division in Korea, is sometimes associated with one region. Nonetheless, Kaufmann's geographic approach should not be pushed too far. Most U.S. combat forces are flexible, as they must be. The United States has too many global allies and interests to create separate force structures to defend each one; the cost of doing so would be prohibitive.

Most U.S. forces are based in the United States and can be deployed to whatever region national command authorities need to send them. Even formations thought of as devoted primarily to one location may, when a crisis erupts, be deployed elsewhere. The Army drew down large numbers of European-focused forces to fight in Vietnam (not to mention in Desert Storm in 1991); of late, even the above-mentioned 2nd infantry division has sent one of the two brigades normally stationed on the Korean penin-sula to Central Command's operations in the Iraq/Afghanistan theater. No combat units had been designated for Afghanistan before 2001. Looking to hypothetical scenarios in the future, there are no U.S. forces dedicated to addressing instability in Pakistan, peacekeeping in Kashmir, humani-tarian relief in Africa or South or Southeast Asia, or a range of other pos-sibilities. Yet capacity must be retained for addressing one or more such scenarios at a time, even if each is individually unlikely to occur.

Kaufmann's framework, while useful, therefore is more notional than precise. It also is not explained in detail in his writings. Kaufmann's per-sonal reputation for rigor and great knowledge in the field makes it likely that his estimates are as reasonable as any other—given the inherent lim-itations of tackling the problem in this basic way—but they are not easily reproducible.

Another way to view it is that even if a specific overseas interest of the United States disappeared or was deemed to require U.S. military protec-tion no longer, the resulting decline in the defense budget would be less than Kaufmann's method suggests. That is because some of the forces that he allocated to a given region are in fact also important for other regions, if not in a primary role then at least as a strategic reserve.

Toward a Bottom Line

So how do we move toward a bottom line in this estimate? It is worth tak-ing a step back to first principles. Since the end of the cold war, the Per-sian Gulf has represented one of two possible areas of operations around which the Pentagon has built its combat force structure, the other being East Asia. Given that the current peacetime defense budget of the United States is just over $500 billion, that might seem to imply costs as great as $150 billion a year to defend the Persian Gulf (factoring out the $200 bil-lion of the defense budget that is devoted to research and development, intelligence, homeland defense, nuclear forces, and other costs that are not easily attributable to any geographic theater, as well as the costs of core military infrastructure including major commands, educational institu-tions, and the like in the United States).

That number is too high, however. The United States has numerous overseas obligations, not just two. Recognizing that fact but still empha-sizing the importance of the Persian Gulf in U.S. military strategy, Bill Kaufmann's 1992 estimate is that the United States spent 20 percent of its peacetime defense budget on defense of the Persian Gulf and the broader Middle East. As noted, applying that same percentage today would suggest a cost of about $100 billion a year (again, not counting the costs of the ongoing Iraq war).

However, as mentioned above, even that estimate is debatable, for several reasons. Most important, many forces that would be assigned to Central Command in wartime are available for other purposes. (This is the obverse of the current situation, in which forces that could otherwise be used in Europe, East Asia, or elsewhere are taking their turn being deployed to Iraq and Afghanistan.) A figure of $100 billion a year may not be a bad estimate of the costs of forces that *most likely* would be deployed to operations in the Persian Gulf, but it overlooks the fact that many of them could have secondary purposes as well.

So Kaufmann's estimate is not a bad starting point, but the most meaningful answer to the question about the bottom line posed above requires one to ask how much *less* the United States would spend on its military overall if the Persian Gulf were somehow dropped from the list of overseas commitments and possible wartime theaters. That is the important policy question, even if others sometimes characterize the problem differently. Answering it requires an effort to estimate what force posture the United States would want to keep. In reality, the savings could be somewhat less than Kaufmann estimated, since some of the forces that normally could be assigned to the Persian Gulf might need to be kept for other possible scenarios (such as stabilizing a collapsing Pakistan or Indonesia; maintaining an air patrol in the Taiwan Strait if China/Taiwan tensions heat up again and remain hot over an extended time; or countering a Russian menace to the Baltic states, now NATO members). This question is too complex and open to interpretation to lend itself to an easy answer—which is one reason why some other scholars, in attempting to view the problem holistically, also have calculated ranges rather than precise estimates of the associated costs.

Estimates in the general range of half of Kaufmann's estimate, or $50 billion a year, present what are probably reasonable answers to the question posed. It is a straightforward proposition to imagine a scenario in East Asia—perhaps another war in Korea or a major multilateral stabilization mission in South Asia or Southeast Asia—that could make demands on U.S. ground forces just as great as those made on forces in Iraq and Afghanistan during this decade. That would mean that ten active-duty Army divisions, three Marine divisions, and associated airpower capabilities might be required even if the Persian Gulf was no longer a theater of any concern to U.S. strategists. To hedge against the possibility of such a scenario, the Pentagon could not scale back its forces quite as much as Kaufmann's method might suggest; in fact, some would surely argue that it could not scale them back at all, though that seems a dubious proposition. There is plenty of room for argument about whether the resulting amount of U.S. military spending on the Persian Gulf should be estimated at $50 billion a year, or 50 percent more or less than that figure, roughly speaking. In any case, that range seems to define the approximate realm of reasonable debate, to the first order.

Notes

1. For a similar view, see Mark A. Delucchi and James J. Murphy, "U.S. Military Expenditures to Protect the Use of Persian Gulf Oil for Motor Vehicles," *Energy Policy* 36 (2008), pp. 2253–64.

2. Ibid., p. 2253.

3. See Under Secretary of Defense (Comptroller), *National Defense Budget Estimates for FY 2008*, pp. 110–27.

4. Found at Office of the Under Secretary of Defense (Comptroller) (www. budget.mil).

5. Raymond Hall, David Mosher, and Michael O'Hanlon, *The START Treaty and Beyond* (Congressional Budget Office, 1991), pp. 62, 135; Office of the Under Secretary of Defense (Comptroller), *National Defense Budget Estimates for FY 2008* (Department of Defense, 2007), pp. 44–52, 81 (www.defenselink.mil/comptroller/ defbudget/fy2008 [accessed April 1, 2007]); and Stephen I. Schwartz, *Atomic Audit: The Costs and Consequences of U.S. Nuclear Weapons since 1940* (Brookings, 1998), p. 3.

6. Naturally, more detailed questions require more detailed analysis. For example, when determining the compensation level needed to ensure a given level of recruiting success, comparison between military and civilian pay levels for individuals of comparable age and skill are needed, as are historical data on the typical correlation between a given pay increase or other improvement in compensation and improved recruiting results. Such comparisons need to go beyond broad metrics to examine various compensation levels and occupational specialties. For example, on balance there is no systematic civilian-military pay gap (especially when all of DoD's compensation, including tax advantages and housing and family subsidies, is counted, without even including retirement or health benefits, which do not accrue to all). Counting all direct benefits, a twenty-two-year-old person with four years' experience and E-4 rank earns the equivalent of $70,000 if unmarried and $85,000 if married with two children. However, certain individuals with specific technical skills may well make substantially less in the military than they would in the private sector. See Carla Tighe Murray, *Evaluating Military Compensation* (Congressional Budget Office, 2007), pp. 22, 31–32;

and Richard L. Fernandez, *What Does the Military 'Pay Gap' Mean?* (Congressional Budget Office, 1999), pp. 1–7.

7. Department of Defense, "Fiscal Year 2009 Budget Request: Briefing Slides," February 4, 2008, p. 5.

8. Jim Garamone, "Bush to Ask for Additional $42 Billion for War Operations," *American Forces Press Service News Articles*, September 26, 2007 (www.defenselink. mil/news/newsarticle.aspx?id=47586 [accessed September 27, 2007]).

9. Department of Defense, "FY 2008 Global War on Terror Request," Department of Defense, February 2007, pp. 11, 72–76 (www.defenselink.mil [accessed March 1, 2007]).

10. Steven M. Kosiak, "The Cost of U.S. Operations in Iraq and Afghanistan and for the War on Terrorism through Fiscal Year 2007 and Beyond," CSBA Update, Center for Strategic and Budgetary Assessments, September 12, 2007, pp. 1–5; and Testimony of Steven M. Kosiak before the Senate Budget Committee, "The Global War on Terror (GWOT): Costs, Cost Growth, and Estimating Funding Requirements" (Washington: Center for Strategic and Budgetary Assessments, February 6, 2007), pp. 1–8.

11. Kosiak, "The Global War on Terror," pp. 3–6.

12. Congressional Budget Office, "Costs of Operation Desert Shield," CBO Staff Memorandum, January 1991, pp. 1–20.

13. Kosiak, "The Global War on Terror," p. 6.

14. David Newman and Jason Wheelock, "Analysis of the Growth in Funding for Operations in Iraq, Afghanistan, and Elsewhere in the War on Terrorism," Congressional Budget Office, February 11, 2008, pp. 1–3.

CHAPTER FOUR

Who's Afraid of China's Oil Companies?

ERICA S. DOWNS

Who's afraid of China's national oil companies? Quite a few people, if the reaction to the unsolicited offer made by China National Offshore Oil Corporation Ltd. (CNOOC Ltd.) for Unocal is any guide. The furor that erupted inside the Beltway in response to CNOOC Ltd.'s bid to break up the merger between Unocal and Chevron highlighted the anxiety that many U.S. policymakers, pundits, and oil companies harbor about the growing global footprint of China's national oil companies (NOCs). The objections raised by opponents of CNOOC Ltd.'s attempted acquisition are rooted in popular perceptions of the Chinese NOCs' international expansion. The conventional wisdom views the NOCs as arms of the Chinese government that are aggressively snapping up exploration and production assets around the world to enhance China's energy security at the expense of that of other consumers. Moreover, it contends that the state financial support that Beijing provides to China's NOCs to achieve this noncommercial objective violates the rules of the game for international mergers and acquisitions because it is not available to Western, publicly traded firms. Consequently, the Chinese government and oil companies are turning the global competition for oil into a game that major international oil companies (IOC) like Chevron cannot even compete in, let alone win.

This chapter is based on Erica S. Downs, "The Fact and Fiction of Sino-African Energy Relations," *China Security* 3, no. 3 (Summer 2007), pp. 42–86.

This chapter examines several popular perceptions about the foreign investments of China's NOCs. Contrary to conventional wisdom, China's NOCs are not merely puppets of the Chinese party-state that are expanding internationally for the sole purpose of assuaging Beijing's concerns about energy security. In addition, the NOCs are not dominating the global exploration and production market or "locking up" oil through their overseas deals and thus denying it to other consumers. State financial support, however, probably does provide China's NOCs with a competitive advantage over other oil companies and may play a larger role in the wake of the financial crisis. Separating myth from reality in the discourse on the foreign investments of China's NOCs is important in order understand whether and to what extent their international mergers and acquisitions impact U.S. interests.

"China's NOCs Are Arms of State Policy."

Not exactly. Conventional wisdom holds that China's NOCs are merely puppets of the Chinese party-state, executing the directives of their political masters in Beijing. As with most conventional wisdom, there is an element of truth in this view. To be sure, the Chinese party-state has several levers of control over the NOCs. However, China's oil majors—with their subsidiaries listed on foreign stock exchanges, global business portfolios, and vast profits earned from the high oil prices of recent years—are powerful and relatively autonomous actors with their own domestic and international interests that do not always coincide with those of the party-state.[1]

China's three major NOCs, China National Petroleum Corporation (CNPC), China Petroleum and Chemical Corporation (Sinopec), and China National Offshore Oil Corporation (CNOOC), grew out of government ministries. CNPC, formed in 1988 from the upstream (exploration and production) assets of the Ministry of Petroleum Industry (MPI), is the biggest oil producer in China and the fifth largest in the world.[2] Sinopec, established in 1983 from the downstream (refining and marketing) assets of MPI and the Ministry of Chemical Industry, has the largest refining capacity in China and the third largest in the world.[3] CNOOC, formed in 1982 as a corporation under the MPI and modeled after Western oil companies, was established to form joint ventures with foreign firms to operate in China's territorial waters and is primarily an upstream company that dominates China's offshore. CNPC and Sinopec are both ministry-level companies, a bureaucratic rank that they fought hard to

TABLE 4-1. Internationally Listed Subsidiaries of China's National Oil Companies

Listed company	Parent company	Percent owned by parent
PetroChina	CNPC	86.29
Sinopec Corp.	Sinopec	75.84
CNOOC Ltd.	CNOOC	66.41

Sources: PetroChina, Form 20-F for the fiscal year ended December 31, 2007, filed with the U.S. Securities and Exchange Commission, p. 80 (www.petrochina.com.cn/resource/EngPdf/annual/20-f_2007.pdf); Sinopec Corp., Form 20-F for the fiscal year ended December 31, 2007, filed with the U.S. Securities and Exchange Commission, p. 59 (http://english.sinopec.com/download_center/reports/2007/20080606/download/Form20F2007.pdf); and CNOOC Ltd., Form 20-F for the fiscal year ended December 31, 2007, filed with the U.S. Securities and Exchange Commission, p. 91 (www.cnoocltd.com/encnoocltd/tzzgx/dqbd/f20f/images/200941157.pdf).

retain during their creation to maintain a privileged position when dealing with the state.[4] CNOOC has the lower status of a general bureau. The current general managers of all three companies—Fu Chengyu (CNOOC), Jiang Jiemin (CNPC) and Su Shulin (Sinopec)—all hold the rank of vice minister. Jiang and Su are also alternate members of the Seventeenth Chinese Communist Party Central Committee, which consists of the 371 most politically powerful individuals in China.

Each of the three companies has a subsidiary listed on the Hong Kong and New York stock exchanges. The parent companies are the majority shareholders of the listed companies (See table 4-1). Other shareholders include individual and institutional investors.

Ownership does not always equal control, and that is true for the party-state. The State Asset Supervision and Administration Commission (SASAC) is the government body with formal authority over China's largest state-owned enterprises (SOEs), including the NOCs. Although SASAC has been relatively passive—it did not collect dividends from its firms until late 2007 and it does not appoint their top leaders (although it does choose high-level managers)—SASAC has begun to exert greater influence over SOEs in recent years by linking managers' salaries to their companies' financial performance.[5] Nonetheless, the party-state primarily controls the NOCs through other sources of influence in the party and the government.

The primary instrument of power that the party-state exercises over China's NOCs is the power to appoint, dismiss, and promote the companies' general managers. The ultimate authority over the top positions in the NOCs rests with the Chinese Communist Party's Organization Department, whose decisions are ratified by the Politburo Standing Committee. This authority extends, indirectly, to the NOCs' internationally listed subsidiaries because an individual appointed general manager of a parent company usually concurrently serves as the chairperson of the board of its listed subsidiary. Consequently, NOC managers must balance corporate and party-state interests, especially if they want to advance their political careers. Executives who demonstrate managerial prowess while not running afoul of the Chinese Communist Party can often use their tenure in the oil patch as a springboard to national leadership.[6]

The party-state also controls the NOCs through its investment approval system. Domestic investments in oil and natural gas fields, pipelines, refineries, oil storage facilities, and liquefied natural gas terminals require government approval. Foreign energy investments in excess of $30 million need to be signed off on by the National Development and Reform Commission (NDRC), and those in excess of $200 million have to be reviewed by the NDRC and then submitted to the State Council for approval.[7]

An additional source of leverage is the provision of cheap credit. In recent years, China's NOCs generally did not require government funds because of their strong cash flows. Nonetheless, low-cost loans from state-owned banks, such as the China Export Import Bank (China Eximbank) and the China Development Bank, can function as carrots and sticks that the party-state can wield over the NOCs.

Influence, however, is a two-way street between the party-state and the NOCs. Indeed, Chinese officials, academics, and journalists have come to view the oil majors as a "monopolistic interest group" that prioritizes profits over social welfare.[8] The Chinese media have criticized China's NOCs for creating artificial oil shortages to pressure the government to increase prices at the pump (discussed below), with one report noting that many people feel that the NOCs are robbing Chinese citizens and the country to bolster their bottom lines.[9] The power and autonomy of China's NOCs is due to a number of factors, including their relative strength vis-à-vis the central government's energy bureaucracy, large profits earned during the recent oil boom, and internationally listed subsidiaries.

The liberalization and decentralization of China's energy sector since the early 1980s, which are part of the broader transition from a centrally planned to a market economy, have shifted power and resources away from the central government toward the state-owned energy companies, notably the NOCs.[10] Multiple bureaucratic restructurings have fragmented

Beijing's authority over the energy sector among many government agencies, some of which are understaffed, underfunded, and politically weaker than the state-owned energy companies. China does not have a single government agency, such as a ministry of energy, with the clout to coordinate the often conflicting interests of the multiple stakeholders.[11] In addition, the transformation of China's energy ministries into corporations resulted in a large transfer of personnel and industry expertise from the government to the companies. Some Chinese analysts describe China's energy sector as one of "strong firms and weak government," with "strong" and "weak" referring to capacity, not authority.[12]

The enormous profits earned by China's NOCs in recent years due to higher oil prices are also a source of clout with the party-state. In 2007, CNPC and Sinopec were the two largest state-owned enterprises by revenue, and the earnings of CNPC alone offset the losses of all loss-making state-owned enterprises.[13] Moreover, among SOEs under the central government in 2007, CNPC, Sinopec, and CNOOC accounted for 24.1 percent of total sales revenue, 23.5 percent of profits, and 40 percent of taxes collected.[14] Although it is difficult to determine how and to what extent profits translate into government influence, some Chinese commentators contend that the companies' contributions to government coffers have bolstered their ability to shape government decisions.[15]

In addition, when CNPC, Sinopec, and CNOOC listed subsidiaries on the New York and Hong Kong stock exchanges in 2000–01, the companies exposed themselves to the influence of actors other than the party-state. These actors include not only the stock exchanges themselves, but also entities such as the U.S. Securities and Exchange Commission, international auditing and engineering firms, independent shareholders, and members of the companies' boards of directors. The independent shareholders of CNOOC Ltd., for example, have compelled the company to take actions counter to its interests and those of its parent company.[16]

China's NOCs sometimes advance corporate interests at the expense of national ones. For example, CNPC and Sinopec have periodically reduced crude runs at their refineries to pressure the government to raise the state-set prices for refined products, which lagged behind the higher crude oil prices of recent years. Their cutbacks created diesel and gasoline shortages in China and prompted the government to raise refined product prices.[17] Similarly, the opposition of China's NOCs is widely cited by Chinese energy experts as one of the main reasons that the Chinese government has not created a ministry of energy, a hot topic of debate in recent years.

The NOCs are reluctant to have another political manager and fear that it would limit their access to China's top leadership.[18] Moreover, the NOCs' acquisition of upstream assets abroad creates diplomatic challenges for Beijing. For example, the pursuit of investment opportunities in Iran by China's oil majors runs counter to the Ministry of Foreign Affairs's objective of curbing Iran's nuclear ambitions. Although the ministry has no direct authority over the NOCs, it has nonetheless pressured them to retreat from Iran, where Sinopec has signed a buyback agreement for the development of the Yadavaran oil field and China's NOCs are negotiating investments in liquefied natural gas (LNG) projects.[19]

"The Energy Security Concerns of the Chinese Government Are Driving the Foreign Investments of China's NOCs."

Yes, but there are also compelling commercial factors fueling the companies' global search for oil. The international expansion of China's NOCs is often portrayed as a misguided attempt by the Chinese government to enhance China's energy security through the acquisition of exploration and production assets abroad. In that view, Chinese leaders are acutely aware that a stable supply of oil is critical to the continued expansion of China's economy, which in turn is necessary for them to remain in power. China's leaders, who believe that oil is "too important to be left to the market" and prefer to "own oil at the wellhead," have dispatched China's NOCs on a global hydrocarbon shopping spree to help satisfy the country's burgeoning demand for oil. To be sure, China's NOCs have a government mandate to supply Chinese consumers with oil and natural gas. However, the tendency of some international observers to portray the foreign investments of China's NOCs as a political project conceived within the walls of Zhongnanhai, the Chinese leadership's compound in Beijing, obscures the market incentives driving the global expansion of China's NOCs.

Reserve Replacement and Diversification

China's NOCs appear to be purchasing exploration and production assets abroad first and foremost to grow and diversify their reserves of oil and natural gas. Like all other oil companies, China's NOCs need to continuously acquire new reserves to replace what they deplete. The opportunities are limited for China's oil companies to substantially grow their reserves, which account for only 1.3 percent of the world's proved oil reserves and 1.1 percent of the world's proved natural gas reserves. Although China's proved reserves of natural gas more than doubled, from 0.89 to 1.88 trillion cubic meters, between 1987 and 2007, China's proved oil reserves declined from 17.4 billion to 15.5 billion barrels over the same period.[20] As a result, overseas assets are important sources of growth in reserves and production for China's NOCs. Indeed, PetroChina's chief financial officer, when discussing his company's first overseas acquisition, noted that "we can hardly expect big production increases at home. Overseas production will become the new driving force in the future."[21]

China's NOCs are also expanding internationally to diversify their reserve portfolios. Like the major IOCs, China's NOCs recognize that it is not smart to put all of their eggs in one single basket. Unlike those of the major IOCs, however, the reserves of China's NOCs are highly concentrated in one country, China. Consequently, China's oil companies are seeking to disperse operational risks by expanding the number of countries in which they have production assets.[22]

Profits

The upstream sector is historically the most profitable part of the oil business. Like the IOCs, China's oil companies seek income from exploration and production assets. Unlike the IOCs, China's NOCs have also sought to raise profits through the expansion of their overseas upstream portfolios to offset losses suffered in their domestic upstream and downstream operations as a result of price controls for crude oil, which were abolished in 1993, and for refined products, which are still in place.

A key driver of CNPC's initial forays abroad in the early 1990s was to recoup some of the money that it was losing through its domestic upstream operations.[23] CNPC had been incurring large losses since its creation in 1988 because the cost of producing a barrel of oil in China was higher than the state-set price for crude oil, at which the company was required to sell the majority of its production. The company hoped to bolster its bottom line by producing oil abroad and selling it on the international market.[24]

In recent years, CNPC and Sinopec have sought to grow their international exploration and production portfolios to help mitigate the heavy losses incurred in their refining operations because state-controlled prices for refined products have prevented the companies from passing on rising crude oil costs to their customers. Between 2001 and 2007, the average annual price of crude oil increased from $26 to $72 per barrel, and China's oil imports grew from 1.6 million to 4.2 million barrels per day.[25] Forced to sell diesel and gasoline below cost, CNPC and Sinopec began to hemorrhage money. Sinopec, which is China's largest refiner and depends on imports for about three-quarters of its crude, suffered the most. The company's billions of dollars in refining losses since 2005—including $8.8 billion in the first half of 2008 alone—have not been completely offset by government subsidies and value-added tax rebates on crude oil imports.[26] Sinopec has sought to partly counter its poor downstream margins through expanded exploration and production at home and abroad.[27]

International Competitiveness

China's NOCs are searching for exploration and development opportunities abroad to transform themselves into world-class energy companies. Their executives recognize that if they want to be internationally competitive, then they must compete internationally. Former CNOOC general manager Wei Liucheng employed a soccer analogy to make that point, arguing that China's oil companies "can't just play in the domestic league. We should also compete in the World Cup."[28]

Some of the overseas assets in which China's NOCs are invested were purchased to gain technical expertise. One objective of CNOOC Ltd.'s bid for Unocal was to gain deepwater exploration and production capacity, while its acquisition of a stake in Canada's MEG Energy was aimed at securing advanced oil sands extraction technology.[29] Similarly, Sinopec, which has the least upstream experience of China's three major NOCs, has sought to enhance its exploration and production expertise through international acquisitions.

China's NOCs are also making international investments to develop the large project management skills possessed by the major IOCs.[30] Companies like ExxonMobil have distinguished themselves by their ability to execute complex projects that involve employing cutting-edge technology, arranging huge financing packages, handling intraconsortium politics and host government relations, managing environmental impacts, and finishing on time and on budget. In contrast, China's NOCs, which are relative latecomers to the international oil business, have less experience in simultaneously managing and coordinating all the components that must come together to execute very large projects overseas. That said, CNPC has gotten its feet wet with the big integrated projects that it operates in Kazakhstan and Sudan. Similarly, Sinopec and CNOOC Ltd. have partnered with IOCs with large project management experience to develop deepwater blocks in Angola (BP-operated Block 18) and Nigeria (Total-operated Oil Mining Lease 130).

FIGURE 4-1. Oil Consumption and Production of Selected Countries, 2007

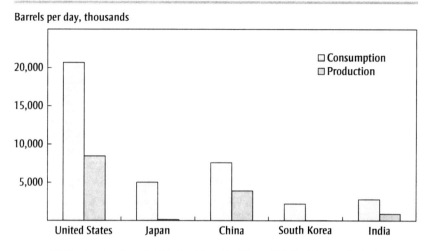

Barrels per day, thousands

Source: U.S. Energy Information Administration, International Energy Statistics, Petroleum Production (http://tonto.eia.doe.gov/cfapps/ipdbproject/IEDIndex3.cfm?tid=5&pid=53&aid=1 [July 17, 2009]); and U.S. Energy Information Administration, International Energy Statistics, Petroleum Consumption (http://tonto.eia. doe.gov/cfapps/ipdbproject/IEDIndex3.cfm?tid=5&pid=54&aid=2 [July 17, 2009]).

Energy Security

China's NOCs are also acquiring assets abroad to help ease the Chinese leadership's concerns about oil supply security. A net oil exporter until 1993, China is now the world's third-largest oil importer, behind the United States and Japan, and the world's second-largest oil consumer, after the United States (figure 4-1). Between 1997 and 2007, China's oil demand almost doubled, from 4.2 million to 7.9 million barrels per day, and the country's oil imports more than quadrupled, from 1 million to 4.2 million barrels per day (figure 4-2).[31] The International Energy Agency projects that by 2030 China's oil demand will rise to 16.6 million barrels per day and its imports will reach 12.5 million barrels per day, making the country dependent on imports for 75 percent of total oil consumption.[32]

Chinese oil executives and senior officials have publicly stated that China's NOCs have a political mandate to enhance China's energy security through investment in foreign oil fields.[33] There is a fairly widespread perception within Beijing that oil pumped by China's NOCs abroad provides a more secure supply of oil than purchases made on the international market. This idea is rooted in skepticism of the view of Western

FIGURE 4-2. China's Oil Demand and Domestic Supply, 1990–2007

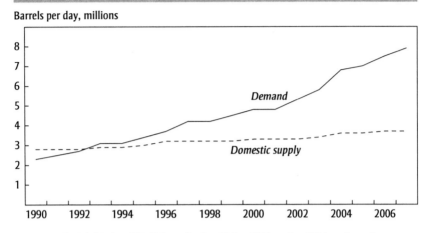

Barrels per day, millions

Source: *BP Statistical Review of World Energy* (London: BP, June 2008), pp. 8 and 11 (www.bp.com).

oil industry analysts, who maintain that the world market will always make oil available to the highest bidder. In the late 1990s, some Chinese energy officials (and at least one Chinese oil company executive trying to gain high-level political support for the international expansion of China's NOCs) argued that China might one day find itself in a situation in which China has money to buy oil but none is available on the international market because of war or other political turmoil.[34] In such a situation, they continued, the Chinese government could order the NOCs to send their foreign oil production back to China. Despite these concerns, it is difficult to imagine a scenario in which China has money but no oil to purchase, because the world is filled with buyers and sellers. Moreover, the NOCs are unlikely ever to pump enough oil abroad to cover China's oil import requirements because more than three-quarters of the world's oil reserves are in countries that do not permit foreign equity participation. Indeed, ExxonMobil, the world's largest "resource-seeking" oil company, pumped only 2.2 million barrels per day overseas in 2007.[35]

"China's NOCs Are Taking Over the World."

No. China's national oil companies are not dominating the international upstream sector. The rapid global expansion of China's NOCs has generated concerns that the Chinese firms are winning the worldwide race for exploration and production assets. Many stories in the mainstream media about the NOCs' expanding global footprint merely list the wide swath of countries in which the companies are invested, giving no information about the size and quality of their assets. Nevertheless, some readers have concluded that China's NOCs have left the IOCs in the dust. The reality, however, is quite different. To be sure, the overseas expansion of the NOCs is certainly changing the competitive landscape of the global oil industry, and some analysts expect the NOCs, especially CNPC, to become international players on a scale to rival that of the major IOCs.[36] However, reports that China's NOCs have already vanquished the competition are exaggerated.

First, China's NOCs have not been as active in global mergers and acquisitions as their international peers, according to a report by UK-based consultancy Wood Mackenzie on the emergence of Asian NOCs in the international upstream sector.[37] To be sure, the value of the acquisitions made by Asia's most expansive NOCs—CNPC, Sinopec, CNOOC Ltd., ONGC of India, and Petronas of Malaysia—grew dramatically, from less than $500 million in 2001 to more than $6 billion in 2005. However, the Asian NOCs' level of participation in international mergers and acquisitions (M&A) during that period still lagged behind that of international companies of comparable scale. The total value of the acquisitions made by the five companies studied over the five years from 2001 to 2005 was $13 billion, compared with $33 billion for BP, ConocoPhillips, ENI, Devon Energy, and Occidental. In the view of Wood Mackenzie, many of the Asian NOCs "have yet to complete deals that reflect the scale of their ambitions in the international upstream sector. This is particularly the case for CNPC, the largest of the Asian NOCs." Wood Mackenzie maintains that CNPC, whose largest foreign purchase was PetroKazakhstan, for which it paid $4.2 billion in 2005, is capable of making acquisitions in the range of $20 to $40 billion.

Moreover, the international M&A activity of China's NOCs slowed considerably in 2007 and 2008.[38] Not only was there stiff competition for assets, but China's oil majors also shied away from major acquisitions to avoid buying at the top of the oil price cycle.[39] They also had some bad luck. CNOOC Ltd., for example, made offers for Shell's assets in Nigeria and Australia but lost out to a local buyer in Nigeria (chosen by Shell to help improve its relations with the country) and to Woodside in Australia (because Woodside had preemption rights).[40]

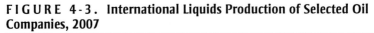

FIGURE 4-3. International Liquids Production of Selected Oil Companies, 2007

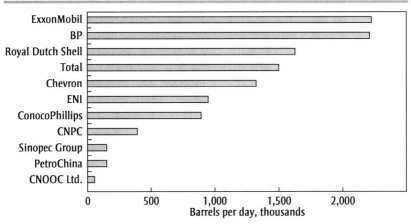

Source: ExxonMobil, *2007 Financial and Operating Review*, p. 60 (www.exxonmobil.com/corporate/files/ news_pub_fo_2007.pdf); BP, *Financial and Operational Information 2003–2007*, p. 57 (www.bp.com/liveassets/ bp_internet/globalbp/ STAGING/global_assets/downloads/F/FOI_2003_2007_full_book.pdf); Royal Dutch Shell, *Financial and Operational Information 2003–2007*, p. 57 (www.faoi.shell.com/2007/ explorationproduction/ oilandgasproduction.php); Total, *Factbook 2000–2007*, p. 38 (www.total.com/static/en/medias/topic2346/ 2007_factbook_global.pdf); Chevron, *2007 Supplement to the Annual Report*, p. 42 (www.chevron.com/ documents/pdf/Chevron2007Annual ReportSupplement.pdf); ENI, *Fact Book 2007*, p. 42 (www.eni.it/ attachments/publications/reports/reports-2007/Fact-Book-2007-eng.pdf); ConocoPhillips, *Fact Book 2007*, p. 3 (www.conocophillips.com/NR/rdonlyres/DC7C811C-4528-4B6F-A6E8-A3D7A172F39B/0/07_Fact_Book.pdf); data provided by Wood Mackenzie to author by e-mail, December 14, 2008.

Second, China's NOCs do not produce as much oil overseas as the major IOCs. In 2007, CNPC and its domestic peers pumped a combined total of 780,000 barrels per day of liquids abroad, less than the overseas production of any of the major IOCs (figure 4-3). Although the NOCs are invested in upstream projects in more than two dozen countries, most of those assets have done little to substantially bolster their overseas output. The foreign production of China's NOCs is concentrated in just two countries, Kazakhstan and Sudan (figure 4-4).

Third, China's NOCs rarely compete head-to-head with the major IOCs. High-profile takeover battles, such as those that pitted CNPC against Texaco and Amoco for Kazakhstan's Aktyubinsk Oil Company; CNOOC Ltd. and Sinopec against the members of the consortium developing Kazakhstan's Kashagan field (ENI, ExxonMobil, Royal Dutch Shell, Total, ConocoPhillips, and Inpex) for British Gas's stake in the project;

FIGURE 4-4. **Overseas Liquids Production of China's NOCs, 2007**
Percent

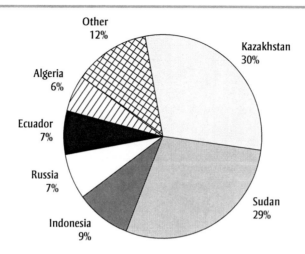

Source: Data provided by Wood Mackenzie to author by e-mail, December 14, 2008.

and CNOOC Ltd. against Chevron for the U.S. firm Unocal have been the exception rather than the rule. Many of the assets purchased by China's NOCs are not especially attractive to the IOCs. During their early forays overseas, the NOCs had little choice but to take what they could get. New to international mergers and acquisitions and eager to secure reserves abroad, the companies largely confined themselves to small projects passed over by the IOCs, whose enormous balance sheets and high cost structures require large projects.[41] The Chinese firms, especially CNPC, accumulated an unwieldy collection of small assets that spanned the globe. While some casual observers in the international media seized on the breadth of the NOCs' portfolios as evidence that Chinese firms were winning the global competition for oil, Chinese industry analysts tended to focus on the lack of depth, bemoaning that the late arrival of China's NOCs to international exploration and production appeared to have doomed them to settling for the "leftovers" of the IOCs. An interlocutor from CNPC lamented that even acquiring the "little bones and little scraps of meat" left behind by the IOCs was difficult.[42]

Although the initial overseas ventures of China's NOCs helped them develop a taste for the substantially bigger assets on which the IOCs feast, the upstream capabilities of the Chinese firms have prevented them from

TABLE 4-2. Selected Large Merger and Acquisition Deals
by China's NOCs

Company	Date	Country	Assets	Price (US$ millions)
Sinopec	Dec 08	Syria	Tanganyika Oil	2,000
CNPC	Nov 08	Iraq	al-Ahdab field[a]	2,900
Sinopec	Jun 08	Australia	AED Oil	561
Sinopec	Dec 07	Iran	Yadavaran field[b]	2,000
Sinopec	Nov 06	Russia	Udmurtneft	3,500
Sinopec	May 06	Angola	Blocks 17 and 18	2,400
CNOOC Ltd.	Jan 06	Nigeria	OML 130	2,300
CNPC	Oct 05	Kazakhstan	PetroKazakhstan	4,000
CNPC/Sinopec	Sep 05	Ecuador	Encana Ecuador	1,400
Sinopec	Mar 05	Angola	Block 18	725
CNOOC Ltd.	Jan 02	Indonesia	Repsol-YPF	585
CNPC	Mar 97	Sudan	Blocks 1, 2, and 4	750

a. Technical service agreement.
b. Buyback agreement, pending final approval from Iran.
Source: Author's database.

directly competing against the IOCs for certain projects.[43] For example, the Chinese oil companies' lack of deepwater exploration and production capacity has limited their ability to bid for some of the most attractive blocks open to foreign investment, which are in deepwater and ultra-deepwater locations in Angola, Brazil, Nigeria, and the United States. Technological constraints have also largely kept China's NOCs on the sidelines of the development of unconventional hydrocarbons and liquefied natural gas.

Faced with those disadvantages, the NOCs have sought to satisfy their appetite for larger assets by investing in countries and projects with elevated levels of political risk, where they face less competition from the IOCs. Many of the largest acquisitions made by China's NOCs are in places where IOCs have been unable or unwilling to tread (table 4-2). Indeed, CNPC has amassed assets worth about $7 billion in Sudan, where the north-south civil war and the violence in Darfur have kept the IOCs away.[44] CNPC and Sinopec, through their joint venture Andes Petroleum, also spent $1.4 billion to purchase EnCana's assets in Ecuador, which the Canadian firm had been trying to divest for more than a year, partly because of the increasingly difficult operating environment for foreign oil companies.[45] CNOOC Ltd.'s tolerance for risk also helped the company gain entry into the Nigerian deepwater—and the opportunity to work with Total and Petrobras—with its purchase for $2.3 billion of a 45 percent working interest in an offshore block (Oil Mining Lease 130) with a controversial and opaque ownership history.[46]

There are two reasons why China's NOCs accept higher levels of political risk than the IOCs. First, China's NOCs have less experience that the IOCs in evaluating political risk because of their substantially shorter involvement in international mergers and acquisitions. Unlike many of the IOCs, China's oil companies have yet to suffer substantial political disruption or expropriation of their overseas operations and therefore attach lower risk premiums to investments in unstable areas.[47] Second, there appears to been a perception within the Chinese oil industry and government, at least during the companies' earlier forays abroad, that Beijing would be able to protect their investments in countries with elevated levels of political risk through its relationships with host governments.[48]

However, China's NOCs are learning that they are not immune to the misfortunes that their IOC peers have suffered in unstable areas. In Ecuador, for example, CNPC and Sinopec have experienced for themselves the difficult operating environment that spurred the exodus of IOCs such as Occidental Petroleum and EnCana. In 2007, the government successfully pressured the Chinese firms to accept less favorable contract terms under the threat of a 99 percent windfall profits tax, resulting in huge losses.[49] In Sudan, where CNPC is operating in fields discovered by Chevron in the 1970s, Chinese oil workers have been kidnapped and killed like their American predecessors. The most recent murders, in October 2008, elicited a public commitment from CNPC to "fully understand the risks of overseas projects."[50]

Moreover, the difficulties that China's NOCs have encountered in Ecuador and Sudan indicate that the Chinese government may do more harm than good when it comes to mitigating political risk. As one Chinese media commentator noted, the fact that Ecuador was on the list of countries in which Beijing encouraged the NOCs to invest in 2007 indicates that the government lacks the ability to assess political risk, let alone the official diplomatic means to protect assets overseas.[51] That point is underscored by CNPC's experience in Sudan, where Beijing's friendly relations with Khartoum have put the company's employees in the crosshairs of various Darfur rebel groups.[52]

FIGURE 4-5. China's Foreign Oil Production in and Imports from Selected Countries, 2007

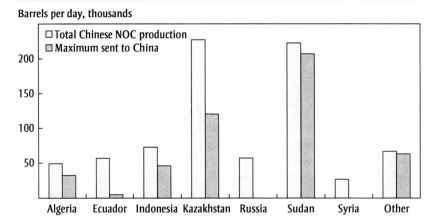

Barrels per day, thousands

Sources: "Table: China Dec. Crude Oil Imports and Exports," *Reuters*, January 21, 2008; data provided by Wood Mackenzie to author by e-mail, December 14, 2008.

"China's NOCs Are Removing Oil from the World Market."

No. The argument that China's oil companies are taking oil off the world market and reducing the amount available to other consumers by selling their overseas oil production exclusively to consumers in China is wrong. Any foreign oil production that China's NOCs send to China merely replaces oil that China would have to buy from other countries. If the NOCs shipped home every one of the 779,000 barrels per day of oil that they produced abroad in 2007 (instead of the maximum of 474,000 barrels per day that they may have sent to China), then China would not have needed to purchase at least 300,000 barrels per day more from other exporters, such as Saudi Arabia and Angola, which are China's top two providers of crude oil and also large suppliers to the United Sates (figure 4-5). Moreover, the NOCs are actually expanding rather than contracting the amount of oil available to other consumers by pumping oil abroad, especially at oil fields in which other companies are unable or unwilling to invest.

In 2007 China's NOCs sold at least 40 percent of their foreign oil production, about 300,000 barrels per day, on the international market. The NOCs did not send home any of the oil that they pumped in Azerbaijan,

Russia, Syria, or Tunisia.[53] Most of the oil produced by China's NOCs in Ecuador was shipped to the United States.[54] At least half of the output of the NOCs in Kazakhstan and one-third of their production in Indonesia was sold locally (figure 4-5).

The export to China of any oil pumped by China's NOCs in Indonesia, Kazakhstan, and Sudan, which accounted for two-thirds of the foreign oil production of the NOCs in 2007, appears to be largely determined by economic factors. China is a natural market for oil from Indonesia and Sudan because of their geographical proximity. Moreover, Indonesia's Minas crude and Sudan's Nile Blend crude, which accounts for the bulk of CNPC's output in Sudan, are very similar to the light and sweet crudes produced in northeastern China and easy for China's refineries to process. Indeed, Indonesia was a large crude oil supplier to China in the 1990s, and China has been the top buyer of Sudanese crude since the country began exporting oil in 1999. (However, the sharp decline in China's oil imports from Sudan in 2006 indicates that the company is happy to sell the oil to consumers in other countries that are willing to pay a higher price than buyers in China.[55]) In addition to sending the bulk of its Nile Blend production to China, CNPC is also importing the Dar Blend crude that it began to pump in Sudan in 2006—and building a refinery to process it—because of the lack of international buyers for this high-acid, heavy-paraffin crude.[56]

In Kazakhstan, CNPC has sold most of its oil production—which is concentrated in the northwestern part of the country—on the international market because it is more profitable to export it to the West through the Caspian Pipeline Consortium or the Aytrau-Samara pipeline than to deliver it to China. Indeed, China's crude imports from Kazakhstan hovered around a mere 25,000 barrels per day until the completion of the easternmost leg of the Kazakhstan-China oil pipeline in 2006. CNPC's production in the Kumkol region, which is located near the mouth of the pipeline, may account for some of the growth in Kazakhstan's oil exports to China, which reached 121,000 barrels per day in 2007 (figure 4-6).

"State Financial Support for China's NOCs Gives Them a Competitive Advantage over the IOCs."

Probably, but it's hard to determine how much of an advantage China's NOCs gain from Beijing's largesse. The Chinese government's willingness to draw on government coffers to help China's NOCs expand internationally has

FIGURE 4-6. China's NOCs' Oil Production in and Imports from Kazakhstan, 1997–2007

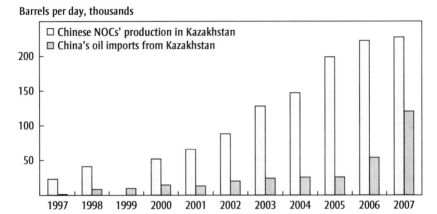

Barrels per day, thousands

Sources: Data provided by Wood Mackenzie to author by e-mail, December 14, 2008, and October 6, 2006, and in person in Beijing, September 2006. Tian Chunrong, "2001 nian Zhongguo shiyou jinchukou zhuangkaung fenxi" [An Analysis of China's Oil Imports and Exports in 2001], *Guoji shiyou jingi* [*International Petroleum Economics*] 3 (2002), p. 12; Tian Chunrong, "2007 nian Zhongguo shiyou jinchukou zhuangkaung fenxi" [An Analysis of China's Oil Imports and Exports in 2007], *Guoji shiyou jingi* [*International Petroleum Economics*] 3 (2008), p. 40.

sounded alarm bells in capital cities and oil companies around the globe. Policymakers and oil executives have raised concerns that Beijing's provision of low-cost capital to the NOCs and development assistance to host countries gives China's oil firms a leg up on the competition for exploration and production assets. To be sure, the Chinese government has given China's NOCs some financial support that is unavailable to the IOCs. However, it is difficult to assess the extent to which Beijing's deep pockets have tilted the playing field in favor of the NOCs because the Chinese firms rarely engage in direct competition with the IOCs. In addition to the dearth of case studies, the waters are further muddied by the ability of China's NOCs to self-finance most of their foreign acquisitions and the fact that the attempts of the Chinese government to use development assistance as a tool to help the NOCs build their international upstream portfolios has yielded mixed results.

The contention that state financial support gives China's oil majors the upper hand in the race for exploration and production assets is difficult to assess because for the most part, China's NOCs and the IOCs compete on different playing fields. There are few examples of acquisitions for which a Chinese NOC and a major IOC engaged in direct competition, and there are even fewer examples for which complete information about the offers made by all bidders is easily accessible. As a result, there is not enough information to make a definitive declaration about how much of an advantage the NOCs derive from state financial support in international mergers and acquisitions.

The financing package that CNOOC Ltd. assembled for its bid for Unocal—one of the rare examples of direct competition between a Chinese oil firm and an IOC for which the information needed to assess the capital costs of both firms was publicly available—lies at the heart of the contention that Chinese government subsidies give China's NOCs a competitive advantage over the IOCs. To finance its $18.4 billion bid for Unocal, CNOOC Ltd. arranged to borrow $7 billion from its parent company, CNOOC, including a $2.5 billion bridge loan with no interest and a $4.5 billion thirty-year loan with a 3.5 percent interest rate. The company also lined up a $6 billion loan from the Industrial and Commercial Bank of China. The fact that the terms of at least some of the loans were not available to any Western publicly traded company prompted Peter Robertson, then the vice chairman of Chevron, whose offer of $16.4 billion had already been accepted by Unocal at the time of CNOOC Ltd.'s bid, to cry foul: "We're not competing with this company; we're competing with the Chinese government—I think it's wrong."[57]

The Unocal case, however, is probably the exception rather than the rule when it comes to Chinese government institutions providing huge sums of cheap capital to bankroll the overseas acquisitions of China's NOCs. As Trevor Houser has noted, Unocal is far and away the largest acquisition ever attempted by a Chinese firm, and it was undertaken by the smallest of China's three major NOCs. Given that the $18.5 billion deal was half of CNOOC Ltd.'s market capitalization and more than double its annual revenue, it is hardly surprising that CNOOC Ltd. had to seek external sources of funding for its offer.[58]

CNOOC Ltd.'s offer notwithstanding, China's NOCs have not bid for assets that are large enough to require huge amounts of external capital. The largest foreign investment made by a Chinese firm, CNPC's purchase of Petrokazakhstan for $4.2 billion, is less than one-quarter of CNOOC Ltd.'s offer for Unocal. Moreover, a number of the international upstream assets recently bought by the NOCs cost $500 million to $2 billion. China's NOCs, which raked in billions of dollars in profits in recent years, were able to self-finance acquisitions in this range easily. Sinopec, for example, stated that it would use internal resources to finance its $2 billion acquisition of Tanganyika Oil in 2008, while CNOOC Ltd. self-financed its $2.3 billion purchase of a stake in Nigeria's offshore Oil Mining Lease 130 in 2006.[59]

While China's NOCs may enjoy a lower cost of capital than the IOCs, the ability of the Chinese firms to accept lower rates of return than the IOCs likely has more to do with lack of shareholder discipline. As Houser observes, firms such as ExxonMobil and Shell undertake projects that they expect will earn returns on reinvested earnings in the mid-to-high teens because they realize that if they deliver anything less, then their shareholders might take their profit as dividend payments to invest in companies that can deliver higher returns. In contrast, the wholly state-owned CNPC, Sinopec, and CNOOC Group are not subject to the same level of shareholder discipline.[60] To be sure, the companies are under increasing pressure from their sole shareholder, the Chinese government, to generate profits. Beijing reinstated the collection of dividend payments from state-owned enterprises in 2007, with the NOCs required to hand over 10 percent of their profits. In addition, the salaries for CEOs of China's "Big Three" oil companies, like those of all other state-owned enterprises under the control of the central government, are now more dependent on their firms' performance than ever before.[61] That said, the Chinese government probably is willing to settle for rates of return that are unacceptable to the shareholders of IOCs because, unlike those shareholders, the Chinese government has objectives other than profit maximization, such as securing access to energy resources abroad.

However, just because China's NOCs are able to live with lower rates of return than the IOCs, they do not necessarily do so across the board. In fact, some recent industry analyses have indicated that the reputation that China's NOCs have earned for paying top dollar for assets is not entirely deserved. A report by Wood Mackenzie concluded that the majority of the deals completed by five Asian NOCs, including the three Chinese majors, over the 2001–04 period would yield rates of return in the range of 15 to 20 percent.[62] Other industry analysts have also noted that the gap between rates of return for Asian NOCs and the IOCs narrowed in recent years as the IOCs increased their oil price assumptions.[63]

The Chinese government's developmental assistance to host countries, like the cheap capital that it offers to China's NOCs, probably provides China's oil majors with a competitive advantage in certain situations over other companies that do not receive similar support from their governments. Anecdotal information indicates that Beijing's attempts to use aid to help China's NOCs acquire upstream assets abroad has yielded mixed results. To be sure, Beijing's financial largesse helped Sinopec move into Angola and probably helped persuade Turkmenistan to award CNPC the first-ever onshore production-sharing agreement (PSA) to a foreign company. In Nigeria, however, arrangements for China's NOCs to obtain upstream assets in exchange for development assistance have failed to win China's NOCs any attractive assets.

Beijing uses development assistance as a tool to further the international expansion of China's NOCs for at least two reasons. First, there is a widespread perception in the Chinese government and oil industry that the NOCs are handicapped in the global competition for oil reserves because they are latecomers to the international oil business. The NOCs have been active abroad only since the early 1990s, while some of the IOCs have been operating overseas for about a century. Their historical experience has given the IOCs a competitive edge that other companies have not been able to replicate.[64] For example, Shell, which entered Nigeria in 1938 and enjoyed a monopoly until the country's independence in 1960, is still the country's largest producer. In the words of CNOOC Ltd. chairman and CEO Fu Chengyu, "it is actually not easy for us to find projects. The oil market already has more than 100 years of history and all of the good projects are already taken. As a newcomer, it is obviously not easy to do well."[65]

Second, the sustained rise in world oil prices from 2002 until mid-2008, like other periods of high prices, shifted bargaining power away from foreign companies and toward resource-holding countries, encouraging them to tighten state ownership and to increase their take vis-à-vis that of foreign firms. Some oil producers in Africa, lacking critical infrastructure and eager to diversify their economies away from oil, sought to capitalize on their newfound positions of strength by linking investments in oil exploration and production to investments in other sectors of the host country's economy.[66] Nigeria, for example, offered preferential rights to oil exploration and production blocks to foreign companies that promise to invest in the country's energy and transportation sectors. Edmund Daukoru, Nigeria's minister of state for petroleum, characterized Nigeria's "oil-for-infrastructure deals" as a tool to spur companies that profit from Nigeria's oil wealth to help develop other sectors of the Nigerian economy. He has criticized the IOCs for failing to provide such assistance: "The best bidders have not helped with our national aspirations. No operator has talked railway to me, no operator has talked shipyard, no operator has talked about generating so much. Nobody has shared our aspirations with us. We are in a hurry to develop. The oil industry has been an enclave industry. We want to break out of the enclave and merge with the greater economy of the country, and we are not getting the response we expect and deserve."[67]

Beijing's deep pockets helped China's NOCs establish a footprint in Angola that they otherwise might not have. If China Eximbank had not extended a $2 billion low-interest loan to Angola in 2004 to finance projects built primarily by Chinese companies, such as the refurbishing of the Benguela Railway, it seems unlikely that Sonangol, the Angolan NOC, would have rejected the deal struck between Shell and India's Oil and Natural Gas Corporation Ltd. (ONGC) for the latter to purchase Shell's 50 percent stake in Block 18 (Greater Plutonio fields) and instead sell it to Sinopec.[68] China Eximbank's largesse may also have contributed to Sonangol's decision to award Block 3/80 to Sinopec after refusing to renew Total's license for it in the wake of the French judicial investigation into alleged arms sales—in breach of international sanctions—made to Angola by businessman Pierre Falcone in the early 1990s.[69]

In Nigeria, however, efforts by Abuja and Beijing to link oil and non-oil investments by Chinese firms have yet to yield any results for China's NOCs. An agreement reached in April 2006 between CNPC and the Nigerian government to allow the company to invest $2 billion in the decrepit Kaduna refinery in exchange for the right of first refusal on four oil blocks in the mini-licensing round in May 2006 fell apart.[70] The four blocks are not especially attractive, and CNPC, after doing some seismic work, decided to relinquish them. CNPC's plans to invest in the Kaduna refinery also were derailed when the Nigerian government sold a 51 percent stake in the refinery to Bluestar Oil, a company run by cronies of former Nigerian president Olusegun Obasanjo, just before he left office.[71] Similarly, an arrangement under which China Eximbank would provide Nigeria with a $2.5 billion loan for the construction of a railroad in western Nigeria and, in return, CNOOC Ltd. would receive the right of first refusal on several oil blocks failed to materialize because of disagreements between CNOOC Ltd. and Abuja over the amount of interest each would pay on the loan.[72]

Chinese aid may also have been a factor in Turkmenistan's decision to sign a production-sharing agreement with CNPC to develop the Bagtiyarlyk field on the right bank of the Amu Darya river, making the Chinese firm the first company to operate onshore in Turkmenistan. China Eximbank extended several hundred million dollars in low-interest loans to Turkmen institutions in 2006 and 2007.[73] That aid, combined with the drilling rigs and assistance provided by CNPC and the company's role in spearheading the development of a pipeline to deliver natural gas from Turkmenistan to China have made CNPC the only foreign oil company allowed onshore in Turkmenistan.[74] CNPC also is providing several billion dollars in funding for the pipeline and building the sections in Uzbekistan and Kazakhstan. The Turkmens have told the major IOCs, such as ExxonMobil, Royal Dutch Shell, and Chevron, which are eager to exploit the country's huge onshore gas reserves, that they will have access only to its riskier, less attractive offshore acreage.[75]

The case studies discussed above indicate that state financial support plays a role in helping China's NOCs acquire exploration and production assets abroad but is not necessarily a decisive factor. With respect to access to cheap capital, the low-cost loans that CNOOC Ltd. arranged to help finance its bid for Unocal certainly made the Chinese firm's offer highly competitive with Chevron's. However, CNOOC Ltd.'s final offer ultimately was not high enough to persuade Unocal's shareholders to terminate their agreement with Chevron. (Although Unocal's former chief executive did say that if CNOOC Ltd. had raised its bid rather than withdrawing it then Unocal would have ended up being acquired by the Chinese firm.[76]). With respect to tied aid, the billions of dollars in low-interest loans that China Eximbank extended to Luanda clearly helped Sinopec acquire some assets in Angola. Yet, as with any cheap capital that Chinese government institutions provide directly to the Chinese oil majors, it is hard to assess how much of an advantage Chinese tied aid gives China's NOCs vis-à-vis the IOCs because it is not clear that the IOCs would bid for some of the assets that the Chinese have pursued in "oil-for-infrastructure" deals if the IOCs had been given the opportunity to do so. While it seems likely that some of the IOCs would have jumped at the chance to compete for a production-sharing agreement for Turkmenistan's Bagtiyarlyk field, it seems unlikely that they would have found the onshore acreage that Nigeria offered CNPC and CNOOC Ltd. to be especially attractive.

Conclusion

The good news for U.S. policymakers and pundits who have been watching the global expansion of China's NOCs with varying levels of anxiety is that several of their concerns about the international mergers and acquisitions of these firms are misplaced. The NOCs are not supplicant arms of state policy, purchasing oil assets abroad for the sole purpose of assuaging the Chinese leadership's concerns about oil supply security. The Chinese oil firms, which must acquire oil and natural gas reserves abroad to help ensure their survival in the oil business, are in the diver's seat when it comes to deciding where to invest and gaining the necessary government approvals. China's NOCs are also not winning the global race for exploration and production assets. Although increasingly internationally competitive, China's oil firms do not dominate the international upstream sector. The lion's share of the world's oil reserves and production is in the hands of state-owned oil companies. Among resource-seeking oil companies, the overseas production of China's NOCs lags behind that of the major IOCs. Moreover, whether the NOCs sell the oil that they pump abroad on the international market or to consumers in China does not affect the amount of oil available to consumers in other countries. Each barrel of overseas production that a Chinese company supplies to China is one barrel less that it must buy on the international market—and vice versa.

Observers of the international activities of China's NOCs should nevertheless continue to pay attention to the provision of state financial support to China's oil majors. Beijing's financial largesse has probably given the NOCs a competitive advantage. However, the extent to which China's NOCs depend on state capital to conduct international mergers and acquisitions and the degree to which such financial support impacts the IOCs have been much less than suggested by CNOOC Ltd.'s offer for Unocal. Not only have China's oil majors been able to self-finance most of their deals, but they also rarely compete directly against the IOCs for assets. That may change, however, with the global financial crisis and lower oil prices. Chinese oil executives and officials view the global economic downturn and oil price drop as providing China's NOCs with a golden opportunity to continue their international expansion because assets are cheaper and there is less competition for them.[77] Although the NOCs, like all other oil companies, have less cash to spend on upstream investments, they can turn to state banks for support. Some banks are willing and able to support the acquisition of oil abroad, as indicated by the more than $44 billion in loans extended by the China Development Bank (CDB) and China Export Import Bank to major energy producers battered by the fall in the price of oil, including Russia, Brazil, Kazakhstan, and Turkmenistan, in the first half of 2009 alone. Although only the loan to Kazakhstan is explicitly linked to acquisition of an upstream asset by a Chinese NOC (CNPC is to acquire a 50 percent stake in Mangistaumunaigas), the Chinese government and China's NOCs undoubtedly hope that these "loans for oil" deals will facilitate upstream investment opportunities for Chinese firms. If Beijing's loans do help China's NOCs win plum assets abroad, such as stakes in Brazil's Santos Basin or Turkmenistan's South Yolotan gas field—both of which are very attractive to the major IOCs—it will be an indicator that Chinese state financial support is tilting the playing field in favor of China's NOCs.

Notes

1. This section draws heavily on Erica S. Downs, "Business Interest Groups in Chinese Politics: The Case of the Oil Companies," in *China's Changing Political Landscape: Prospects for Democracy*, edited by Cheng Li (Brookings, 2008), pp. 121–41.

2. "PIW's Top 50: How the Firms Stack Up," *Petroleum Intelligence Weekly*, December 1, 2008, special supplement, p. 2.

3. Ibid.

4. Susan Shirk, *The Political Logic of Economic Reform* (University of California Press, 1993), p. 94.

5. For more information on SASAC, see Barry Naughton, "Profiting the SASAC Way," *China Economic Quarterly*, June 2008, pp. 19–26; Barry Naughton, "SASAC and Rising Corporate Power in China," *China Leadership Monitor* 24 (Spring 2008); and Barry Naughton, "Claiming Profit for the State: SASAC and the Capital Management Budget," *China Leadership Monitor* 18 (Spring 2006).

6. For a discussion of how former NOC general managers Ma Fucai (CNPC), Li Yizhong (Sinopec), and Wei Liucheng (CNOOC) fared in balancing corporate and party-state objectives, see Downs, "Business Interest Groups in Chinese Politics."

7. National Development and Reform Commission, Order 21, "Jingwai touzi xiangmu hezhun zanxing guanli banfa" [Temporary Regulations for Overseas Investment Project Approval], October 9, 2004 (http://tzs.ndrc.gov.cn/tzfg/hjbaz/t2005 1010_44800.htm).

8. "Zhongshiyou wei shenme gan 'zuo gei lianghui kan?'" [Why Did CNPC Make a Gesture of Deference to the Two Sessions?], *Xinhua*, March 12, 2007 (news.xinhuanet.com/legal/2007-03/12/content_5833290.htm [May 15 2007]); and "Zhengxieweiyuan fansi qunian 'youhuang'" [A CPPC Member Reflects on Last Year's 'Oil Shortage'], *Jinghua shibao* [*Beijing Times*], March 13, 2007, p. A4.

9. Shu Xingxiang, "Youhuang 'bigang' zhangjia he bu hefa?" [Oil Shortage 'Forces' a Price Increase; Is It Legal?], *Zhongguo Qingnian bao* [*China Youth Daily*], May 22, 2009 (http://finance.people.com.cn/GB/9345250.html); and "Longduan qiye xianqi 'shehui zeren re' qiye jiaoqu baixing bumanyi" [Monopolized Enterprises Causing Calls for 'Social Responsibilities'; Enterprises Say Accusations Unfair, Citizens Are Not Satisfied], *Zhongguo Caijing Bao* [*China Financial and Economic News*], September 10, 2009 (http://finance.people.com.cn/GB/7882502.html).

10. J. Sinton and others, *Evaluation of China's Energy Strategy Options*, LBNL-56609, May 16, 2005, p. 4 (http://china.lbl.gov/publications/nesp.pdf).

11. For more information on China's fractured energy bureaucracy, see Edward A. Cunningham, "China's Energy Governance: Perception and Reality," Audit of Conventional Wisdom Series, MIT Center for International Studies, March 2007 (http://web.mit.edu/cis/pdf/Audit_03_07_Cunningham.pdf); Kong Bo, "Institutional Insecurity," *China Security* (Summer 2006), pp. 65–89; and R. Lester and E. Steinfeld, "China's Energy Policy: Is Anybody Really Calling the Shots?" Working Paper Series, Industrial Performance Center, MIT-IPC-06-002 (Massachusetts Institute of Technology, January 2006).

12. Huang Hui, "Guojia nengyuan lingdao xiaozu zhi shi di yi bu" [The National Energy Leading Group Is Only the First Step], *Liaowang xinwen zhoukan* [*Liaowang News Weekly*], no. 23 (June 6, 2005), pp. 40–42.

13. Naughton, "Profiting the SASAC Way," pp. 24–25.

14. Chen Geng, "Shiyou gongye gaige kaifang 30 nian huigu yu sikao" [A Review of and Reflections on the 30 Years of Reform and Opening Up of the Oil Industry], *Renmin Ribao* [*People's Daily*], November 18, 2008 (http://energy.people.com.cn/GB/8356135.html.)

15. Interview, Beijing, April 25, 2007.

16. Laura Santini, "Moving the Market: Shareholders Say 'No' in China—CNOOC Is Defeated on Cash Lending; Will Action Spread?" *Wall Street Journal*, April 3, 2007; and Wendy Lim and Tony Munroe, "Update 2–CNOOC Shareholders Reject Proposals on Acquisitions," *Reuters*, January 2, 2006.

17. Trevor Houser, "The Roots of Chinese Oil Investment Abroad," *Asia Policy* 5 (January 2008), p. 152.

18. See the remarks of Zhou Dadi in "Liu Keyu: Zujian guojia nengyuanbu tiaojian yijing jiben chengshu" [Liu Keyu: The Conditions for Establishing a National Ministry of Energy Are Already Basically Mature], *Jingji cankao bao* [*Economic Reference News*], January 29, 2008 (http://news.xinhuanet.com/fortune/2008-01/29/content_7516573.htm).

19. Author's interview with Chinese foreign policy expert, Beijing, June 24, 2008.

20. *BP Statistical Review of World Energy* (London: BP, June 2008), pp. 6 and 22.

21. Xie Ye, "PetroChina Adopts Global Aim," *China Daily*, March 7, 2003.

22. For a Chinese oil executive's perspective on the importance of geographic diversification, see Qiu Zilei, "Zhonghaiyou weihe shougou haiwai zichan" [Why CNOOC Purchases Overseas Assets], in *Gei caijing jizhe jiangke* [*Talking to Financial Reporters*] (Beijing: CITIC Publishing House, 2004), pp. 129–44.

23. Interviews with current and former employees of China's NOCs in Beijing in May 2000, March 2003, and April 2006.

24. Neither Sinopec nor CNOOC suffered from the tightly controlled crude oil prices to the extent that CNPC did. Sinopec had no upstream operations at the time and benefited from a fixed spread for refining margins. CNOOC, which was never subject to price controls or production quotas, exported most of its output prior to the liberalization of crude oil prices in 1993.

25. The prices referenced are spot West Texas intermediate prices. *BP Statistical Review of World Energy*, June 2008, pp. 8, 11, and 16.

26. "Sinopec Gets $4.4B Handout," *International Oil Daily*, July 31, 2008.

27. "Profile: Sinopec Takes the Lead," *Energy Compass*, October 3, 2008; and Chen Wenxian, "Sinopec Aims to Achieve More," *Xinhua China Oil, Gas, and Petrochemicals*, May 14, 2007.

28. "2001 niandu jingji renwu—Wei Liucheng" [2001 Economic Person of the Year—Wei Liucheng], *Duihua* (Dialogue), January 7, 2002 (http://cctv.com/financial/dialogue/sanji/sanji_nr020107_01.html).

29. Nelson D. Schwartz and Doris Burke, "Why China Scares Big Oil," *Fortune*, July 25, 2005.

30. This paragraph is based on e-mail correspondence with Mikkal Herberg on December 12, 2008; a colleague at an international oil company on December 5, 2008; and a Beijing-based energy consultant on December 14, 2008.

31. *BP Statistical Review of World Energy*, pp. 11 and 20.

32. International Energy Agency, *World Energy Outlook 2008* (Paris: OECD/IEA, 2008), pp. 93, 102.

33. See, for example, the remarks of Sinopec Corporation president Wang Tianpu in Duan Xiaoyan, "Women de liyi zhuyao bushi laizi longduan" [Our Profits Mainly Are Not from Monopoly], *21 shiji jingji baodao* [*21st Century Business Herald*], January 8, 2007 [www.nanfangdaily.com.cn/jj/20070108/zh/200701080009.asp]; and the reported remarks of Zeng Peiyan in "Zhongshiyou: dang mengxiang zhaojin xianshi" [PetroChina: When Dreams Become Reality], *Jinrongjie wang* [*Financial World Online*], February 14, 2008 (http://stock1.jrj.com.cn/news/2008-02-14/000003285143.html).

34. See Zhang Yuqing, "Shishi shiyou gongye 'zou chu qu' de fazhan zhanlüe" [Implement the 'Go Abroad' Development Strategy of the Oil Industry), *Hongguan jingji guanli* [*Macroeconomic Management*] 10 (2000), pp. 5–6; and "Guojia jiwei jiaonengsi fusizhang Xu Dingming tan Woguo jingwai shiyou ziyuan de kantan kaifa" [Xu Dingming, Deputy head of the Transportation and Energy Department of the State Planning Commission, Discusses China's Overseas Oil Exploration and Development], *Zhongguo jingji daobao* [*China Economic Herald*], October 8, 1997; and Zhao Yining and Pu Shouru, "Zhongguo shiyou mianlin de tiaozhan" [The Challenges China's Oil Faces], *Liaowang xinwen zhoukan [Liaowang News Weekly]* 9 (1997), p. 13.

35. ExxonMobil Corporation, *2007 Financial and Operating Review*, p. 60.

36. Wood Mackenzie, "The Impact of Asian NOCs on the Upstream M&A Market," *Corporate Insights*, May 2006.

37. This paragraph is based on Mackenzie, "The Impact of Asian NOCs on the Upstream M&A Market."

38. Song Yen Ling, "China: Challenging Deals," *Energy Compass*, July 25, 2008; and "China Spending on Oil M&A Tumbles, but for How Long?" *Dow Jones Energy Service*, November 11, 2007.

39. E-mail from Beijing-based industry analyst, January 14, 2009. See also Kong Di, "Zhongshihua: 'san banfu' yingdui weiji" [Sinopec: 'Three Weapons' to Deal with the Crisis], *Zhongguo qiye bao* [*China Enterprise News*], January 16, 2009 (www.ceccen.com/index.php?page=news_view&id=262).

40. E-mail from Beijing-based journalist, December 14, 2008.

41. Mikkal E. Herberg, "Energy Security Survey 2007: The Rise of Asia's National Oil Companies," NBR Special Report 14 (December 2007), p. 23.

42. Huang Zhouhui, "Xuannian diechu: Zhongshiyou shougou PK you neiqing" [Twist after Twist: The Inside Story of CNPC's Acquisition of PetroKazakhstan],

Minying jingji bao [*Private Economy News*], October 26, 2005 (http://biz.ixco.com/htmlnews/2005/11/01/703774.htm).

43. This discussion of the connection between the upstream capabilities of China's NOCs and their investments in countries with elevated levels of political risk is informed by Houser, "The Roots of Chinese Oil Investment Abroad," pp. 155–58.

44. Estimate based on data provided by Wood Mackenzie in April 2007.

45. Bill Graveland and James Stevenson, "EnCana Sells Ecuador Assets to Chinese Joint Venture for $1.4B US," *Ottawa Citizen*, September 14, 2005; and Claudia Cattaneo, "EnCana Exits Ecuador with US$1.4B Deal: 'Ecuador Is a Difficult Area Both Politically and Economically for the Oil Industry,' " *National Post's Financial Post & FP Investing*, September 14, 2005.

46. Iyobosa Uwugiaren, "Nigeria: Obasanjo's Messy Oil Deals—$20 Billion Investment at Risk," *All Africa*, December 31, 2008; and "CNOOC to Buy Nigeria Block for $2.27 Bil; Block Holding Akpo Field Drew Broad Interest," *Platts Oilgram News*, January 10, 2006.

47. Houser, "The Roots of Chinese Oil Investment Abroad," p. 157.

48. An interlocutor from Sinopec, for example, speaking in 2004 about the operations of China's NOCs in Sudan, stated that political risk was a problem to be solved through diplomacy. See "Ruzhu Sudan shinian Zhongguo shiyou haiwai zhanlue nanbu jiu lu" [After Ten Years in Sudan, It Is Hard for CNPC's Overseas Strategy to Follow the Same Old Road], *Zhongguo jingying bao* [*China Business News*], August 20, 2004 (http://auto.sohu.com/20040820/n221632951.shtml).

49. Wang Kangpeng, "Zhongguo Eguaduoer jiang qian 10 yi Mei yuan daikuan huan shiyou huiyi" [China and Ecuador Will Sign a US$1 billion Loans-for-Oil Agreement], *Renmin Wang* [*People's Daily Online*], April 8, 2009 (http://energy.people.com.cn/GB/9090632.html); "Ecuador Pres Shrugs Off China Arbitration Threat over Oil Tax," *Dow Jones International News*, November 22, 2007. See also "Andes Petroleum Signs Contract Amendments with Ecuador Govt," *Dow Jones International News*, August 26, 2008.

50. Li Xiangyang, Liu Hongbo and Li Bin, "Kuaguo yingjiu: weile jiu wei gurou tongbao" [An International Rescue Effort for Nine Compatriots], *Zhongguo Shiyou bao* [*China Petroleum News*], November 6, 2008 (http://news.cnpc.com.cn/system/2008/11/06/001207678.shtml).

51. Yu Liang, "Bei wusu yu bei sunhai de Zhongguo nengyuan zhanlue" [A Humiliated and Undermined Chinese Energy Strategy], *Zonghe zhoukan* [*Far and Wide Journal*], November 2007 (www.fawjournal.com/archives/399). See also He Huidong, "Haiwai xun you zaoyu dongdaoguo zhengce bianlian" [The Search for Oil Overseas Runs Up against a Host Country's Hostile Policy Change], *Xin Caijing* [*New Finance*] 1 (2008) (http://review.ec.com.cn/article/spfms/spwzjx/200801/542619_1.html).

52. Mohamed Osman, "Darfur Rebels Attack Chinese-Run Oil Field in Attempt to Broaden War against Sudan Government," *Associated Press Financial Wire*, October 25, 2007.

53. Data provided by Wood Mackenzie, December 15, 2008; and Nelli Sharushkina, "China Has Little to Show for Udmurtneft Purchase," *Nefte Compass*, December 13, 2007.

54. Lisa Viscidi, "China: The Ecuador Connection," *Energy Compass*, July 6, 2007.

55. For more on the sale of Sudanese crude to Japan, see Houser, "The Roots of Chinese Oil Investments Abroad," pp. 162–63; and Arthur Kroeber and G.A.

Donovan, "Sudan Oil: Where Does it Go?" *China Economic Quarterly*, Quarter 2 (2007), p. 18.

56. European Coalition on Oil in Sudan, "Sudan's Oil Industry: Facts and Analysis," April 16, 2008, p. 30 (www.ecosonline.org/back/pdf_reports/2008/dossier%20final%20groot%20web.pdf).

57. James Politi, "CNOOC Funding for Unocal Scrutinized," *Financial Times*, June 29, 2005.

58. Houser, "The Roots of Chinese Oil Investment Abroad," p. 159.

59. Song Yen Ling, "Profile: Sinopec Takes the Lead," *Energy Compass*, October 3, 2008; and CNOOC Ltd., "Discloseable Transaction Relations to Acquisition of Interests in Offshore Nigerian Mining License," February 20, 2006, p. 5 (www.cnoocltd.com/UploadFile/NewsFile/b00f5301-d23f-49fa-bcef-c4d8b290dc82.pdf).

60. Houser, "The Roots of Chinese Oil Investment Abroad," p. 159.

61. Naughton, "Profting the SASAC Way," pp. 19–26.

62. Wood Mackenzie, "The Impact of Asian NOCs on the Upstream M&A Market."

63. James Batty, "Corporate: NOC Deals: Fact and Fiction," *Energy Compass*, May 23, 2008.

64. I thank Edward Morse for this point.

65. "Fu Chengyu: 'shenxin zouchuqu' de zhengzhi jingjixue" [Fu Chengyu: The Political Economy of 'Cautiously Going Abroad'], *21 Shiji jingji daobao* [*21st Century Business Herald*], December 29, 2004 (www.nanfangdaily.com.cn/southnews/zt/2004nztk/21ren/200412290045.asp).

66. "Nigeria to Offer New Blocks in Late February; Indian Companies Expected to Bid and to Move Downstream," *Platts Oilgram News*, January 17, 2007; and "Angola, Nigeria Seek Development as India Eyes More Oil Blocks," *Indo-Asian News Service*, September 30, 2005.

67. Tom Ashby, "Nigeria Ties Korea, Taiwan Oil Deals on Bidding Eve," *Reuters News*, August 26, 2005.

68. For discussions of the link between Chinese aid to Angola and Sinopec's acquisition of Shell's 50 percent stake in Block 18, see Xu Fei, "Shiyou daqiao, Zhongguo qiye de Angela wubu" [Building a Bridge to Oil, the Angolan Dance of Chinese Companies], *Nanfengchuang*, November 20, 2006; Margaret McQuaile, "Africa is Staging Ground for Push by China, India to Control More Output," *Platts Oilgram News*, October 26, 2004. For more on China's financial relationship with Angola, see Indira Campos and Alex Vines, "Angola and China: A Pragmatic Partnership," Center for Strategic and International Studies, June 4, 2008, pp. 3–7.

69. "Angola Carries Out Earlier Threats to Total over French Arms Allegations," *World Markets Analysis*, November 8, 2004.

70. This deal was part of a broader memorandum of understanding signed by President Hu for China to provide billions of dollars for investment in Nigerian infrastructure. Information about the disintegration of the "package deal" involving CNPC is based on e-mail correspondence from a Beijing-based oil analyst, July 1, 2007.

71. CNPC made a low bid for the 51 percent stake in the Kaduna refinery in the May 2007 auction ($102 million versus the winning bid of $160 million offered by Bluestar Oil) because CNPC had been told in advance that it would not win. E-mail correspondence with Beijing-based oil industry analyst, July 1, 2007.

72. David Winning, "China CNOOC Deal for Four Nigeria Oil Blocks Hits Snags," *Dow Jones International News*, November 2, 2006.

73. The largest of these loans was a twenty-year loan for $300 million with a 3 percent interest rate extended to Turkmenistan's State Bank of Foreign Economic Activity for work on a fertilizer plant and construction of a glass factory. See "Zhongguo zhengfu jingmao biaotuan fangwen Tukumansitan qude youanman chenggong" [The Visit of the Chinese Government's Economic and Trade Delegation to Turkmenistan Achieves Complete Success], Economic and Commercial Counselor's Office of the Embassy of the People's Republic of China in Turkmenistan, August 29, 2006 (http://tm.mofcom.gov.cn/aarticle/jmxw/200608/20060803009073.html).

74. E-mail correspondence from oil industry analyst, January 15, 2009.

75. Paul Sampson, "Turkmenistan: Looking East," *Energy Compass*, December 5, 2008; and "Turkmenistan Reserves: Big, but Nasty," *World Gas Intelligence*, November 26, 2008.

76. Gary Gentile, "Former Unocal CEO Says Higher Bid Would Have Led to Acquisition by CNOOC," *Associated Press Worldstream*, October 13, 2005.

77. For the comments of Chinese oil executives and officials on the opportunities provided by the financial crisis and lower oil prices, see Kong Ji, "Zhongshihua: 'san banfu' yingdui weiji" [Sinopec: 'Three Weapons' to Deal with the Crisis], *Zhongguo qiye bao* [China *Enterprise News*], January 16, 2009 (www.ceccen.com/index. php?page=news_view&id=262); Zhang Guobao, "Dangqian de nengyuan xingshi: 'wei' zhongzhi 'ji' " ['Opportunities' amid 'Risks' in the Current Energy Situation], *Renmin ribao* [*People's Daily*], December 29, 2008 (http://scitech.people.com.cn/GB/8591346.html); and "Jinring wei ji he youjia baodie gei Woguo youqi haiwai binggou chuangzao liangji" [The Financial Crisis and the Slump in Oil Prices Creates a Good Opportunity for International Acquisitions by China's Oil Firms], *Xinhua*, November 5, 2008 (http://news.xinhuanet.com/energy/2008-11/05/content_110309781.htm).

PART II

Understanding Energy Interdependence

CHAPTER FIVE

Making Sense of "Energy Independence"

PIETRO S. NIVOLA with ERIN E. R. CARTER

S ome of us are old enough to remember Richard M. Nixon proclaiming that "our national goal" should be "to meet our own energy needs without depending on any foreign sources."[1] All of us, old and young, ought to be startled that, thirty-five years later, it remains hard to find a leading U.S. politician who does not champion more or less the same strange notion. Regrettably, that has included two of the nation's most sensible political leaders, President Barack Obama and Senator John McCain. Both of their campaigns repeatedly lamented the nation's "dependency" on foreign oil.

One purpose of a presidential campaign is to win the White House, but another is to educate the public and prepare it for the policy challenges ahead. The 2008 election was uplifting in many respects, but alas, its treatment of the energy issue was not among them. For all the persistent political fascination with "energy independence," the reasoning behind it is flawed. Policymakers from the top down ought to recognize that reality and start leveling with the voters about it.

The aim of this chapter is to encourage a long-overdue change in the terms of what has otherwise become a repetitious and largely sterile debate.

FIGURE 5-1. Crude Oil Prices in the United States and United Kingdom

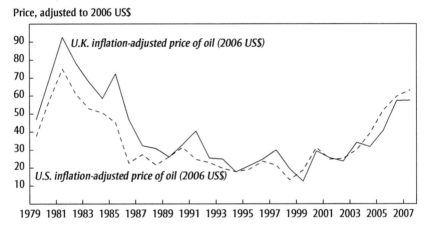

Price, adjusted to 2006 US$

Sources: U.S. price of oil from Department of Energy, table 9.1, Crude Oil Price Summary (http://tonto.eia. doe.gov/merquery/mer_data.asp?table=T09.01); U.K. price of crude oil from Energy Information Administration, table 11.7, "Crude Oil Prices by Selectd Types, 1970–2009" (http://www.eia.doe.gov/emeu/aer/txt/ptb1107.html).

Reality Check

What's wrong with the premise that energy autarky is a path to national prosperity and security? To begin with, the assumption seems to be that the less oil that the United States buys from abroad, the more insulated the U.S. economy will be from the vagaries of the international oil market. By that logic, presumably, if the country imported little or no oil, it would not experience the price fluctuations that it must endure by being too dependent on imports.

A simple way to shatter that myth is to compare the pattern of prices of crude oil in the United States, which has to buy a lot of foreign oil, with the pattern in, say, the United Kingdom, a nation that has been self-sufficient in oil since 1980. Figure 5-1 displays the comparison. The ups and downs of prices in the two countries follow much the same paths. So, for example, when global oil prices (adjusted for inflation) rose sharply, from $25 per barrel in 2000 to more than $66 per barrel in 2007, British consumers were no better insulated from the increase than Americans were. Both faced approximately the same conditions. Petroleum is priced in a world market and no country, even a net exporter, can stop the world and get off.

F I G U R E 5 - 2 . Price of Oil and Changes in U.S. Real GDP

Growth rate   Price, inflation-adjusted 2006 US$

Sources: Price of oil from Department of Energy, table 9.1, Crude Oil Price Summary (http://tonto.eia.doe. gov/merquery/mer_data.asp?table=T09.01); GDP data from Bureau of Economic Analysis (www.bea.gov/ national/xls/gdpchg.xls).

That's the first point to make about the quixotic quest for energy independence. Here is a second: although the U.S. economy today has to import about 60 percent of the oil that it consumes, it is actually *less,* not more, sensitive to rising international oil prices now than it was in Nixon's day, when imported oil amounted to only a third of U.S. consumption.

If you have a hard time believing that, consider figure 5-2, which shows the relationship between movements in oil prices and U.S. rates of economic growth. Following the first energy crisis—the price shock that followed the Arab oil embargo in 1973—the United States fell into recession. When prices skyrocketed with the Iranian revolution in 1979–80, U.S. growth plunged sharply again. The same effect occurred, albeit less markedly, after oil prices ticked up around the time of the Gulf war in 1990. After that, however, an intriguing thing happened: sharp new spikes, like the great run-up starting in 1998, evidently took much less of a toll on the economy. In fact, growth in the four years from 2003 through 2006 was relatively solid despite soaring oil prices. The economic decline that began afterward had less to do with those prices than with the subprime mortgage debacle and the ensuing meltdown in financial markets.

The U.S. economy's sensitivity to energy shocks has diminished because a nation's so-called energy intensity, not the share of fuel supplied by foreign sources, determines its relative capacity to minimize damage from surging energy prices. To produce a dollar of GDP, the United States requires about 40 percent less energy than it did some twenty-five years ago. With energy inputs now playing a proportionately smaller part in generating overall output, the economy absorbs higher fuel prices more easily.

The inflationary (and then contractive) impacts of energy-price hikes, in short, seem to have subsided over time. In any event, sound monetary management and a further reduction in energy intensity are more promising approaches to ensuring economic stability than a struggle to curb dependence on imports.

"Energy Security"

Proponents of energy independence, however, advance additional rationales. One is that by substituting domestically produced fuels for oil from overseas, the United States could help improve the global supply, thereby dampening the world price. That proposition rests on the fact that the United States consumes about a quarter of the world's oil, so, *ceteris paribus*, any appreciable U.S. reduction would transform the international marketplace.

The trouble with that thesis is that only in utopia can *ceteris* be *paribus*. In the real world, other big consumers keep emerging, and they will erase much of the slack that the United States could conceivably cut. Think about China. The gap between Chinese and U.S. GDP is projected to narrow dramatically by 2027. China, which consumes 7.6 million barrels of oil each day, could be on track to add another 3.5 million barrels a day to worldwide demand by 2017. Such an increase alone would offset more than three-quarters of the 4 million barrels a day that the administration of President George W. Bush had proposed to displace by a combination of conservation and use of alternative fuels. With China, India, and other huge new customers coming on line, schemes like Bush's Twenty in Ten Plan might shift the projected global demand for oil to a lower trend line, but it would still climb at an impressive rate. In sum, even if the United States kicked its "addiction to oil"—lowering consumption by 20 percent, as the Bush program had envisioned, and then holding it flat at approximately Europe's expected level—*worldwide* demand would nonetheless resume rising robustly once the global economy recovers from its current downturn.[2]

FIGURE 5-3. Global Forecast of Oil Consumption[a]

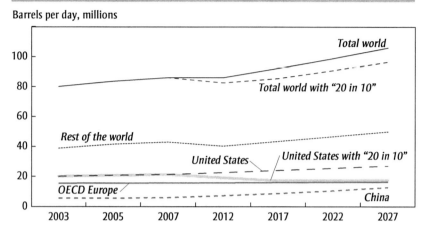

Barrels per day, millions

Sources: Energy Iniformation Administration, www.eia.doe.gov/oiaf/ieo/pdf/ieoreftab_4.pdf;
www.eia.doe.gov/emeu/steo/pub/3atab.pdf; www.eia.doe.gov/emeu/international/RecentPetroleum
ConsumptionBarrelsperDay.xls.

a. Using the EIA's estimation of the average growth rate for the period 2003–30 for each country or area, we interpolated values for 2012, 2017, 2022, and 2027. We know of no published estimate for the U.S. rate of growth after 2017 under the "20 in 10" regime. Therefore, we assumed conservatively that the growth rate would correspond to the OECD's estimate for Europe in the period 2017–27.

Figure 5-3 tells the tale. If anything, this figure paints a best-case scenario. Not only does it assume, heroically, that all of the reduction proposed by a plan such as Bush's would actually transpire—and that subsequently the United States would become almost as energy thrifty as Europe—it also projects conservative growth of demand in China. It presupposes that although China's GDP will close in on U.S. GDP by around 2027, Chinese oil consumption will still be less than 60 percent of the U.S. level—a debatable forecast.

When confronted with this disagreeable reality, the proponents of energy independence repair to yet another argument: granted, whatever energy measures the nation takes will eventually be dwarfed by global demand, but at least, as then-senator Hillary Clinton explained as she prepared to enter the presidential primaries, it would become somewhat less "dependent on regimes that are going to undermine our security."[3] Presumably, the likes of Iran, Venezuela, and Sudan would exert less leverage in international affairs if their oil revenues declined. The United States could help cut these derelicts down to size by lessening the U.S. footprint in the market for their oil.

FIGURE 5-4. Total U.S. Imports of Crude Oil and Products by Country of Origin, 2007

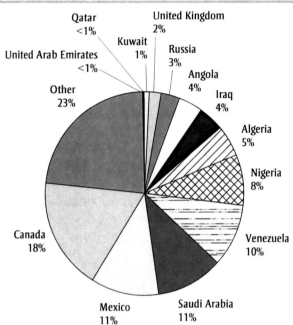

Source: Energy Information Administration, "U.S. Imports by Country of Origin, Total Crude Oil and Products, 1960–2007" (http://tonto.eia.doe.gov/dnav/pet/pet_move_impcus_a2_nus_ep00_im0_mbbl_a.htm).

Would that matters were so simple. The offending oil regimes will enrich themselves whether the United States does business with them or not. Iran, for example, has not sold a single barrel of oil to the United States since the hostage crisis in 1979, yet the mullahs continue to rake in money from the oil that they sell to Europe, Japan, China, and other major clients.[4] The result? Tehran remains defiant, disagreeable, and emboldened. So much for the United States reaping any geostrategic advantage by abstaining from Iranian oil.

Likewise, as figure 5-4 indicates, the United States purchases no oil from the rogue regime in Sudan. However, the Chinese, among others, buy plenty.[5] So long as the Sudanese can peddle their petroleum to *somebody,* U.S. policymakers will remain just as powerless to slow the flow of revenue to that country—and just as wobbly in mobilizing the international community over the atrocities in Darfur—as they would be if the United States were one of Khartoum's direct customers.

Figure 5-4 does show that a nontrivial share of the oil that the United States imports—10 percent—comes from Venezuela. The coffers of Hugo Chávez are being filled, to an extent, by U.S. petrodollars. But suppose that the practice ended tomorrow morning. Venezuela would promptly sell its oil somewhere else, and Chávez would continue to be, well, Chávez.

At the end of the day, the bilateral shopping decisions of the United States matter less than is widely assumed in the vast global energy market. All the main suppliers have plenty of other greedy buyers waiting in line. Yes, there could be trouble if one or more of the foreign sources abruptly interrupted its flow of supplies not just to us but the rest of the world. The price of oil would zoom again. It is impossible to rule out a crisis of that sort. A devastating terrorist attack on Saudi Arabian oil fields, for example, could precipitate it; so could a willful decision by a country with an oddball ruler, bent on wrecking everyone's economy including his own. It is worth noting, though, that even the likes of Ahmadinejad and Chávez show no signs of pursuing a course so masochistic. The fortunes of their regimes depend on pumping oil, not hoarding it. Whatever the case, this much is clear: the effects of a disruption would be felt in the United States, like everywhere else, regardless of whether we are part of a particular supplier's clientele or not.

In sum, it is far from clear how much security is likely to be achieved by becoming more self-sufficient. Now, let us consider the other side of the ledger: what we stand to lose.

The Cost of Cobbling at Home

None of the skepticism expressed so far would be fatal if the search for independence had a minimal economic downside. Unfortunately, the added cost of relying increasingly on homemade fuels is large.

First, a few fundamentals. Seldom acknowledged amid the rhetoric in political circles is that, in point of fact, the United States of America produces the bulk of the "energy" that it needs. True, imports of oil have increased (mostly because Americans choose to drive far more—and use much less efficient motor vehicles—than do consumers in other industrial countries), but imported oil is just one part of the picture. Nearly all of what propels the nation's electric generators—coal, gas, nuclear power, hydropower, and nonhydro renewables—is made in the U.S.A. In stark contrast to western Europe, for example, the United States produces about

85 percent of its primary heating fuel, natural gas, domestically (almost all the rest comes from Canada). Furnaces in Europe had to shut down when the flow of Russian gas through Ukraine was interrupted last winter. Nothing comparable threatens U.S. households.

When we are told that "we must reduce our dependence on foreign sources of *energy*," what does that mean? Is 70 percent self-sufficient too little? Is 80 percent the magic number, or 90 percent, or 100 percent? More likely, the intended point of the statement is that the United States purchases too much foreign *oil*. But even that proposition gets tossed about carelessly. Glance again at figure 5-4, which delineates the various sources of the oil that the country imports. The portfolio is very diversified. Nearly 90 percent of total usage is met by U.S. wells and those of suppliers outside the Middle East. Both of our NAFTA trading partners supply us with more oil than Chávez's Venezuela does, and they supply more than Saudi Arabia does as well.

The unstable Middle East—that is, Saudi Arabia and other Persian Gulf countries—meets less than 11 percent of U.S. needs.[6] Whether that share is thought of as large or, all things considered, comparatively small (Japan, by contrast, imports nearly 90 percent of its oil from the region), importing some Middle Eastern oil makes eminent economic sense. Saudi Arabia and other Gulf countries hold a comparative advantage: they are the world's lowest-cost producers. Not to purchase at least a portion of U.S. crude oil inventory from them—and instead contort ourselves to displace their oil with homegrown fuels—would be a little like me deciding to cobble my own shoes instead of "importing" them from a shoe store.

It is wasteful to insist on fabricating in-house commodities that the United States is better off acquiring through international trade. To cease trading for oil, including with various Middle Eastern sources, would be inefficient, diverting scarce resources into fundamentally non-competitive enterprises and leaving fewer for other industries—industries that could put those resources to more productive use. In the end, living standards are lowered, not secured, by a monomaniacal pursuit of energy independence.

Pause to ponder the U.S. ethanol industry. Some 3.9 billion gallons of ethanol were produced from corn in 2005. That is a drop in the bucket; it amounts to less than 3 percent of total gasoline sales. Many prominent politicians want to jack that quantity way up, but what those enthusiasts don't advertise is the cost.[7]

Ethanol's energy content is appreciably less than that of ordinary gasoline: you need 1.5 gallons of ethanol to drive the same distance you could on a gallon of gasoline. And, if you allow for the substantial subsidy in effect since 2004, ethanol's expense clearly exceeds that of conventional gasoline. In the spring of 2006, for instance, the wholesale price of gasoline was about $2.20 a gallon. The price of ethanol, counting the subsidy, was more like $3.16 a gallon.[8] In some states, the figure was much more.

Some of that price premium might be worth footing if production of corn-based ethanol yielded a significant net reduction of greenhouse gases. Sadly, that is not the case. The lower emissions of carbon dioxide obtained by substituting corn for oil as the feedstock for motor fuel are largely offset by additional emissions of other pollutants, such as nitrous oxide, a potent greenhouse gas. (Nitrous oxide is a byproduct of the nitrogen fertilizer used to grow the corn.) Further, when ethanol refineries are coal-fired (witness the brand-new big plant in Richardson, North Dakota), their effect is to increase, not reduce, carbon emissions.[9]

Because relying on corn-based ethanol promises, at best, only a minor mitigation of greenhouse gas pollution, biofuel advocates are exploring alternatives to corn—fuel derived from cellulose, for instance, or from soybeans, or switchgrass, or even an odd mixture of prairie vegetation. A team of ecologists at the University of Minnesota claims that an eclectic assortment of prairie grasses could offer a bigger environmental payoff; the root structure of this biofuel source, these experts say, acts like an efficient carbon sink.

We don't know enough about the economics of every imaginable substitute for corn, but we do have indications about some. The cost of producing ethanol from cellulose, for example, currently surpasses that of producing traditional ethanol, to say nothing of ordinary gasoline (see figure 5-5). Spending so much for options like fuel from cellulose or soybeans—and consequently crowding out extensive acreage used for food production—can be justified only if the resources thereby diverted are really being allocated to their most-valued use. It strains credulity to claim that they are. Think of it this way: doggedly devoting vast swaths of food-producing farmland to supply motor fuels in effect presupposes that society values filling the tanks of SUVs with the derivatives of grains, beans, or other agricultural commodities more than ensuring affordable grocery bills for hundreds of millions of human beings.[10]

Finally, the pursuit of such perverse priorities also has wide-ranging political implications: it inspires a host of other lobbies to assert that they,

FIGURE 5-5. Cost of Production for Transportation Fuels, 2005

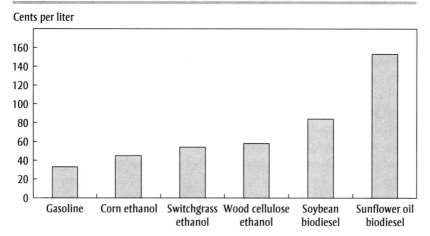

Source: David Pimentel and Tad W. Patzek, "Ethanol Production Using Corn, Switchgrass, and Wood; Biodiesel Production Using Soybean and Sunflower," *Natural Resources Research* 14, no. 1 (March 2005).

like the ethanol coalition, have a legitimate part to play in the great energy-independence game.

The Energy Pork Barrel

To succeed without outsized subsidies, government policies to encourage homemade energy would have to be buttressed by a prolonged period of steep market prices for fossil fuels and also by a long suspension of politics as usual. Don't bet the ranch that either of those conditions will prevail.

Recall the original "National Energy Policy" that President Bush advanced early in his presidency. The ink on it was barely dry when market prices shifted unexpectedly. Natural gas, which had run to $10 per 1,000 cubic feet in early 2001, was closer to $3 by that summer, and, for a while, it was headed lower. Crude oil prices plunged from about $30 a barrel in early September 2001 to around $17 a barrel by mid-November. Prices softened everywhere, including in California, where spot prices of electricity had soared during the state's power crisis in the first part of 2001. Suddenly it looked as if Bush's plan, replete as it was with incentives to goose energy production, had run into the headwind of market forces. The plan's financial practicality, as well as its urgency, quickly faded. Presently, it looks as though a similar train of events—the collapse of oil prices that began in the latter part of 2008—will strand new proposals like those that more recently captivated Congress. Whenever energy prices tumble, as they have repeatedly, the government's latest best-laid plans are bound to follow yesterday's into the boondoggle bin.

The caprice of the marketplace frustrates energy planning. So does the fact that legislative efforts to promote energy self-reliance are perforce politicized.[11] During the troubled 1970s, the Carter administration mounted the most concerted and sustained campaign to enact a national energy plan. Scrambling to create the coalitions needed to pass a bundle of bills, Congress heard from almost all comers seeking a piece of the action. The queue of claimants included energy producers of nearly every shape and description but also other supposed stakeholders.[12] Truckers, for instance, lobbied for extra incentives to pay for windscreens on cabs and trailers. The intercity busline industry sought to get its axles greased with tax benefits, on the grounds that using buses conserves oil. Barge operators on waterways lobbied to secure their own tax preferences, arguing that they, like the buses, were energy savers. Even opponents of school integration got into the act: they labored to graft anti-busing amendments onto bills on the theory that those measures, too, would spare fuel. In the end, to be sure, not every sort of supplicant got its appetite satisfied. The prospect of federal subsidies and dispensations, however, had clearly invited a feeding frenzy by interest groups—and many would keep circling Washington for decades.

Thus, a quarter-century later, the pursuit of energy independence (or, for that matter, energy "security") remains vulnerable to similar political manipulation. Legislation before Congress in recent years has illustrated the extensive logrolling involved. H.R. 3221 was stuffed with loans, loan guarantees, grants, procurement mandates, or tax advantages for small businesses, green-building retrofitters, railroads, bicyclists, and electric vehicle manufacturers, as well as ethanol plants and planters, biodiesel producers, renewable energy manufacturers, developers of hydrogen technology, and nuclear power.[13]

Figure 5-5 provides estimates of the costs of at least some of these alternative energy sources. They are uneconomic in comparison with conventional sources. At a time when the government is running up colossal deficits, devoting large sums of money to prop up costly homespun alternatives for fossil fuels requires, at a minimum, a more compelling justification than just the mantra of "energy independence."

Bottom Line

Neither the economic nor the security interests of the United States are likely to be well served by any national energy strategy that force-feeds, in effect, a "Buy American" approach when international trade can meet a sizable share of U.S. energy requirements at lower cost. Time and again, events have vindicated this conclusion.

Does that mean that there is no reason whatsoever to rework the nation's energy policies? A serious effort to lower the country's level of carbon emissions—not just from the combustion of petroleum products but, also important, coal—is worthwhile, especially if it could encourage other big polluters (China, for example) to participate in a global assault on greenhouse gases. That is because climate change—unlike "foreign oil"—indisputably is a problem, one beckoning for every major economy to take action.[14]

The actions required to combat climate change, however, are rather different from the policy biases that have long dominated Congress's energy agenda. Throwing more tax dollars at ethanol production or pinning hopes on flawed conservation programs, such as U.S. corporate average fuel economy (CAFE) regulations for motor vehicles, are not steps in the right direction.[15] Instead, a serious energy initiative to slow global warming would include a genuine inducement to curb the burning of all fuels that warm the Earth's atmosphere.

To elaborate in depth on how a proper inducement could be designed and implemented would exceed the scope of these pages. Suffice it say that *real* change would begin by adopting a carbon tax—and tying this fundamental reform to a comprehensive overhaul of the nation's anachronistic system of taxation, which perversely penalizes earning, saving, and investment instead of discouraging profligate consumption of energy resources, especially the sorts that are endangering the planet.

Notes

1. Richard M. Nixon, launching Project Independence in November 1973, quoted in John L. Moore, *Continuing Energy Crisis in America* (Washington: CQ Press, 1975), p. 2.

2. President Bush used the "addiction to oil" phrase in his January 2006 State of the Union address.

3. Remarks, 2006 *Take Back America Conference*, June 14, 2006.

4. Japan, for example, imported 684,300 barrels daily from Iran in 2006. See OPEC, "Annual Statistical Bulletin 2006" (www.opec.org/library/Annual%20 Statistical%20Bulletin/pdf/ASB2006.pdf).

5. See, for instance, Peter S. Goodman, "China Invests Heavily in Sudan's Oil Industry," *Washington Post,* December 23, 2004, p. A1.

6. The mix from this region includes oil from Bahrain, Iraq, Kuwait, Qatar, Saudi Arabia, and the United Arab Emirates.

7. In June 2007, the Senate passed an energy bill mandating the production of 36 billion gallons of ethanol per year by 2022. How that target can be reached remains a mystery. Even if *all* U.S. corn production were turned into ethanol, it would yield about 28 billion gallons, still well short of the mandated 36 billion. Let's suppose, for the sake of argument, that "only" half of all corn production were dedicated to ethanol, thereby providing 14 billion gallons. Further let's suppose that the remaining 22 billion gallons were supplied by somehow cranking up production of other sources of ethanol (alcohol derived from cornstalks, wood chips, or grass). To produce 22 billion gallons of it each year, some 25 million acres of cropland would have to be set aside. What might so large a corn-based and cellulosic conversion ensure? Thirty-six billion gallons of ethanol a year would (because of ethanol's lower heat content) displace a mere 1.54 million barrels of oil per day, or 7.4 percent of the 21 million barrels of oil per day the United States *currently* consumes. Of course, by 2022, U.S. consumption is expected to reach nearly 25 million barrels per day, so the dent would be even less impressive. See Robert Bryce, "The Great Corn Con," *Slate,* June 26, 2007. On projected U.S. consumption (for 2020), see Energy Information Administration (www.eia.doe.gov/oiaf/ieo/pdf/ieoreftab_4.pdf).

8. Michael B. McElroy, "The Ethanol Illusion," *Harvard Magazine,* November–December, 2006.

9. There are, of course, other environmental complications of ethanol production. Not too long ago, during a debate at the University of Iowa, for example, it was noted that "when you put a gallon of Iowa ethanol" into your car, you are "also effectively putting several pounds of Iowa topsoil into the Mississippi River." Further, "while growing corn for ethanol production in Iowa can be done without irrigation, doing so in the dryer Plains states cannot. Using modern methods, it takes about 2,000 gallons of water to grow a bushel of corn." Unsustainable depletion of water tables was a problem in the Plains states even before the ethanol campaign took hold. See Michael O'Hanlon, "Beyond Corn-Based Ethanol," *Washington Times,* October 26, 2007.

10. The recent government-engineered surge in ethanol production devours 20 percent of the U.S. corn crop—and raises the price of everything from cereals to feed for chickens, pigs, and dairy and beef cattle. Marianne Lavelle and Bret Schulte, "Is Ethanol the Answer?" *U.S. News and World Report,* February 12, 2007, p. 34.

11. We are not alone in observing this propensity. Vijay Vaitheeswaran, for example, writes: "Modern history is littered with failed government schemes to reduce dependence on fossil fuels, such as the effort to develop synthetic petroleum after the 1970s oil shocks or the false dawn of wind and solar energy in the 1980s. Massive government programs are simply not the answer. Of course, there is a legitimate government role in investing in and encouraging long-term energy research of the sort that private firms cannot provide. But when bureaucrats start picking pet technologies,

whether they are fuel cells, clean coal, or corn-based ethanol, the real trouble begins." "Oil," *Foreign Policy,* November-December 2007, p. 30.

12. See Pietro S. Nivola, *The Politics of Energy Conservation* (Brookings, 1986), p. 7.

13. See Congressional Research Service, "Omnibus Energy Efficiency and Renewable Energy Legislation: A Side-by-Side Comparison of Major Provisions in House-Passed H.R. 3221 and Senate Passed H.R. 6," September 4, 2007.

14. For a good general summary, see Gregg Easterbrook, "Case Closed: The Debate about Global Warming Is Over," *Issue in Governance Studies,* Brookings, June 2006.

15. On the inefficiency of CAFE regulations, see Pietro S. Nivola and Robert W. Crandall, *The Extra Mile: Rethinking Energy Policy for Automotive Transportation* (Brookings, 1995). The main drawbacks identified by this analysis, written a dozen years ago, are inherent in the regulatory program and therefore persist. CAFE standards slowly alter the fuel economy of *new* vehicles, not of the fleets already on the road, and they do nothing to change the driving habits (vehicle miles traveled) of motorists.

CHAPTER SIX

Energy Security
Call for a Broader Agenda

JONATHAN ELKIND

When he ran for office eight years ago, George W. Bush pledged to usher in a new day in U.S. energy security. He pilloried the outgoing Clinton-Gore administration for allowing an energy crisis to take shape. He promised to reduce reliance on foreign oil and complained that the United States had no comprehensive energy policy.[1] More than eight years later, with the new administration of President Barack Obama in office, the problems facing the United States in relation to energy are no less challenging. Global oil prices reached new all-time highs in the summer of 2008 before dropping precipitously as a consequence of the global financial crisis. Analysts are debating whether prices will rebound abruptly or gently, in 2009 or later. OPEC has pushed through production limits in an attempt to firm up oil prices, even as producer and consumer countries continue trading theories about a truly sustainable price and trading accusations about whose policies caused the most recent spike in prices.

Projections of future energy challenges dwarf the headaches that countries are dealing with today. If previous growth trends in energy demand return after the pause brought on by the global financial crisis, countries will struggle to raise the capital required to build the energy infrastructure that they need to sustain their economic growth—especially the emerging economies of the developing world. All countries will struggle that much more to raise the additional capital required to respond to climate change. The latter task will require technological breakthroughs on a massive scale, and it will require a willingness on the part of many to alter the routines of daily life—always a sensitive matter.

The question of the moment is whether President Obama and other newly arrived U.S. political leaders will perform better than their predecessors. Voters called for change in the 2008 presidential and congressional elections, but the kind of change required with respect to energy security will be especially challenging for two reasons: The change will—*must*—deeply affect the way that people live their lives. And the change will not occur simply because it is declared to be overdue. Policymakers will need to tailor signals that guide the marketplace efficiently toward a low-carbon future. But even as they pursue that agenda—stimulating innovation, fostering development of new technologies, discouraging energy consumption, and sustaining investment in the fuels and infrastructure that will create the bridge to the future energy economy—they must avoid picking specific technological winners and must play both a long and a short game at once.

The United States needs to enhance its energy security in a manner that simultaneously deals with near-term risks and protects the global climate. It needs to provide clarity to encourage and then sustain badly needed—and massive—investments. It needs to take its proper role in responding to one of the existential issues of our times. This agenda requires maturity and foresight if we are to succeed where progress has eluded us for so long.

Call for a Broader Vision of Energy Security

For more than thirty years, the United States has struggled to enhance its energy security. Unfortunately, its efforts have been episodic rather than systematic, and they have focused on a narrow definition of the term "energy security"—one that excludes environmental sustainability. U.S. attention to energy security typically reaches a fever pitch when global energy prices spike or international conflict threatens to disrupt energy trade. The nation typically—and vainly—responds by seeking a technological or legislative "silver bullet." None exists. Its half measures do not fundamentally alter energy consumption or supply patterns. The attention paid to the issue by the public and policymakers alike wanes as soon as prices subside naturally, which they generally tend to do in a sector that is predisposed to long, recurring business cycles. The lower prices lull the country into a false sense of security even as the energy intensity of its economy remains substantially undiminished. In a few years, the cycle repeats.

Meanwhile, over the past twenty years, the scientific community has spoken with a steadily greater consensus: human-caused greenhouse gas emissions are changing the global climate. Human beings are placing natural systems—and over time, large numbers of human lives—at risk. The threats now faced by human systems are unprecedented, and many of the people who face the most severe impacts of a changing climate will be those least able to protect themselves.

Certain details of the climate system remain elusive, and the range of politically and economically practicable policy responses is the source of at times fiery debate. Also unclear are the precise technological solutions that will allow nations to provide the energy services that their populations require and sustain their economic growth while radically reducing greenhouse gas emissions. As a consequence, some people throw up their hands and wait for better or more palatable prescriptions; others grasp at any possible policy, missing the point that the very scale of global climate change requires policies and measures that are effective, efficient, and replicable.

The linked energy and climate challenges that the United States faces require a more aggressive, more thorough, more sustained response than the country has mustered so far. New threats to the functioning of international energy markets have taken shape—threats ranging from rapidly growing competitors for traditional hydrocarbon resources, to terrorists whose willingness to wreak human suffering and economic chaos is beyond debate, to environmental impacts that threaten the global climate system. It is high time for the country to engage in a more concerted, serious effort. The good news is that there are several obvious areas where it can focus its efforts and begin to make early progress.

Elements of Energy Security

Traditional definitions of energy security have included availability, reliability, and affordability.[2] Clearly a contemporary understanding of energy security must include those three dimensions, but now it must also include a fourth—environmental sustainability. (Table 6-1 summarizes the elements of energy security.)

Availability

First and foremost, energy security stems from the availability of energy goods and services—consumers' ability to secure the energy that they need.

TABLE 6-1. Energy Security: Elements, Components, and Potential Threats

Elements	Components	Potential threats
Availability	Physical endowment of producer-countries	Exhaustion of reserves that can be extracted cost effectively
	Ability of producers, transit countries, and consumers to agree on terms of trade	Limits on development opportunities (such as resource-nationalist policies and state-to-state contracts)
	Technological solutions for production, transportation, conversion, storage, and distribution	Problems in siting infrastructure—for example, the "not in my back yard" (NIMBY) syndrome
	Capital investment	Financial, legal, regulatory, or policy environments that are not conducive to sustained investment
	Viable legal and regulatory structures	
	Compliance with environmental and other regulatory requirements	
Reliability	Robust, diversified energy value chain	Failure of energy systems due to severe weather, earthquake, and so forth
	Adequate reserve capacity for entire value chain	Failure due to poor maintenance or underinvestment
	Short- and long-term protection from terrorist attacks, extreme weather, and political interruptions	Attack (or threat of attack) by military forces or terrorist organizations
	Adequate information about functioning of the global energy market	Political interventions (such as embargoes and sanctions)
Affordability	Low price volatility	Exhaustion of reserves that can be extracted cost-effectively
	Transparent pricing	Excessive demand resulting from high energy intensity and/or failure to institute sound pricing and other desirable policies
	Realistic expectations for future price—affordability is not simply a matter of absolute cost of energy but also a matter of expected future price compared with current price	Failure to incorporate an environmental dimension into concepts of energy security, resulting in need for an even more urgent response to climate change or other threats to sustainability
	Prices that reflect full costs, as a matter of short-term incremental cost and over the full lifecycle	
Sustainability	Low emissions of greenhouse gas and other pollutants	Policy responses to narrow definition of energy security (for example, support for increased use of coal before carbon capture and storage technologies are commercialized)
	Minimal contribution to local, regional, or global threats to environmental quality	Impacts of a changing climate (such as sea-level rise, storm surges, and severe weather events)
	Protection of energy systems from impacts of a changing climate	

Source: Author's compilation.

Availability requires the existence of commercial energy markets in which buyers and sellers trade energy goods and services, markets that take shape only when parties agree on terms that accommodate the commercial, economic, political, strategic, and other interests of buyers, sellers, and shippers. Mutuality of interest among the players in the value chain is therefore a prerequisite for energy security. Nonetheless, the extent of the power of individual players in the marketplace and the skill with which those players pursue their individual interests will determine the extent to which the terms of trade favor one party or another. Creation of energy markets requires physical resources, capital investment, the efficient application of technology, proper legal and regulatory frameworks, products that comply with legal and regulatory requirements, and societal acceptance of the given energy service. As this listing makes clear, the idea of "availability" is not quite as simple as it may seem at first glance.

Over recent decades, demand for energy has skyrocketed across the globe due to sustained economic growth in industrialized countries and accelerated growth in China, India, and other emerging economic powers. The transportation sector, which depends heavily on petroleum-based fuels, has expanded especially rapidly in the latter countries. Past oil and gas development has depleted the relatively easy-to-access petroleum reserves; because of that, future oil and gas development will involve deposits that generally are

—scarcer, with fewer super-giant fields being discovered

—farther from existing demand centers

—deeper and harder to extract, often involving deepwater locations, high pressure, or high sulfur content

—located in poorer countries, with risks of political instability and poor governance

—concentrated in areas where governments restrict access, whether as a matter of cartel membership, as in OPEC countries, or in response to other policy priorities, such as environmental concerns

—costlier to develop.

Many energy systems require literally tens of billions of dollars of investment and a decade or more of planning and construction before they begin operations. Such systems will not come into existence unless the entire set of prerequisite factors is aligned. And, as discussed later, ensuring alignment of all those factors in a time of major evolution in energy markets will present major challenges.

Certain energy resources may be abundant but commercially unavailable due to technology gaps. One such example is the class of compounds called methane hydrates, in which methane is held in lattice form under high pressure at low temperatures. Methane hydrates have the potential to add substantially to global natural gas reserves if the methane can be extracted on a cost-effective basis. Unfortunately, even after decades of research, widespread commercial exploitation of methane hydrates is not imminent.[3]

Other energy resources may be available using current technology, but their extraction would conflict with other economic activities or, as in the case of the Arctic National Wildlife Refuge (ANWR), environmental policies. Huge political controversy often results. During the summer of 2008, both U.S. presidential candidates—Senator Barack Obama and Senator John McCain—joined with President Bush in calling for the opening of the outer continental shelf (OCS) for drilling along parts of the U.S. coast long unavailable for petroleum development. All of the Pacific coastline, all of the Atlantic seaboard, and the Florida coast of the Gulf of Mexico have been essentially off limits to oil and gas development since Union Oil's platform A blew out in 1969, soiling beaches and pleasure craft in Santa Barbara and other prime locations in Southern California. Since that time, the environmental performance of the petroleum industry has taken huge strides forward, but even when international energy companies are struggling to find accessible reserves for development, as they are today, most of the outer continental shelf has remained off limits.

Reliability

Reliability involves the extent to which energy services are protected from interruption. Energy is an essential building block of economic activity; it enables daily life. Interruptions jeopardize the ability to run factories, illuminate hospitals, and heat homes continuously. In certain cases, therefore, energy reliability can be a matter of life and limb. Ways to enhance energy reliability include the following:

—diversifying sources of supply
—diversifying the supply chain used for processing, transporting, and distributing energy
—increasing the reserve capacity of energy networks such as pipelines and power generation and transmission systems

—reducing energy demand, which can ease the burden on overstretched distribution infrastructure
   —creating emergency stocks
   —developing a redundant infrastructure
   —disseminating timely market information.

Several of these points deserve special attention. First, energy security is a much broader and more comprehensive notion than some of the ideas that are frequently used as shorthand in political debates. For example, independence from foreign oil, a reflexive pursuit of the U.S. political system since the time of the OPEC embargoes in the 1970s, is not the same as energy security. To see why, one only need look at the challenges that faced the United States after hurricanes Katrina and Rita hit the coast of the Gulf of Mexico in late summer 2005. The reliability of U.S. energy supplies was temporarily undermined due to problems affecting *domestic* supply lines. More than 100 oil and gas production platforms in the gulf were damaged by the storms, and nearly 20 percent of the country's refinery capacity was taken out of service. The country's only deepwater oil import facility, the Louisiana Offshore Oil Port (the LOOP) was also disabled for a time after each of the storms.[4] The country experienced energy security challenges regardless of the fact that the interrupted parts of the supply chain were completely within U.S. control. Access to imported fuel was actually critical to restoring a reliable energy supply inside the U.S. market after Katrina and Rita.

It is certainly true that reducing dependence on foreign oil imports might be desirable for other policy reasons. For example, current U.S. consumption of foreign oil contributes to a massive trade imbalance because the nation transfers hundreds of billions of dollars overseas. Moreover, as has been noted pointedly since September 11, 2001, some of those oil payments result in revenues for parties that sponsor terrorist groups.[5] Nonetheless, even if the United States had zero dependence on foreign oil, it would face energy security challenges as a result of its considerable energy intensity. Domestically produced oil—or economic substitutes for oil—would be just as subject to price fluctuations in an integrated global oil market as current imports are. In times of high global prices, U.S. producers of petroleum or ethanol would be tempted to export production, which in extremis could squeeze supply for domestic consumers.

A second point worth additional scrutiny is the notion of redundancy, reserve capacity, and emergency stocks. As noted above, many parts of the energy economy are exceptionally capital intensive. That is certainly the case if one is talking about transmission lines, fuel stocks, and extra capacity for power generation or fuel refining. A key challenge, therefore, is determining who pays for the planned redundancy. As a member of the International Energy Agency, the United States is required to maintain a minimum of ninety days of net crude oil imports in reserve in case of a large-scale interruption of global supply; the oil, paid for by the U.S. government, is to be sold on the market during times of crisis.[6] At present, the United States holds just over 700 million barrels in the Strategic Petroleum Reserve (SPR), which is equivalent to about fifty days of net imports. In his State of the Union address in January 2007, President Bush declared his intention to double the size of the SPR, to 1.5 billion barrels. It is much more costly to maintain strategic stocks of natural gas than of crude oil because of the technical complexity of storing sufficient volumes of natural gas.

With respect to surplus operating margins for the electricity industry, if one wishes utility companies to build and maintain extra generating capacity and transmission and distribution (T&D) systems, one needs to account for the additional capital expenditure in the rate base so that investors can earn a fair return on their investment. Failure to provide the proper legal and regulatory framework to encourage the construction and reliable operation of T&D systems can lead to disastrous outcomes. The California electricity crisis of 2001, for example, occurred when legislators embarked on a deregulatory initiative that resulted in a mismatch of power demand and the T&D capacity required.

A final point goes to the importance of information. Many aspects of contemporary energy markets are truly global. Demand or supply dynamics in one corner of the world trigger reactions—price volatility, fuel switching, capital investment choices—far away.[7] For that reason, information—especially information about energy prices—is essential for ensuring the reliability of energy systems.[8] A disruption of oil production in Nigeria—such as has occurred repeatedly over the past few years—has an impact on oil prices everywhere. An accident at a nuclear plant outside Tokyo results in a spike in demand for alternative power-generating fuels like liquefied natural gas (LNG), which causes a rise in prices for LNG shipments in both the Pacific Basin and the Atlantic.

Affordability

Energy that is not affordable in absolute terms is energy that cannot be used, and in fact roughly 1.8 billion people worldwide suffer chronically from what is sometimes referred to as energy poverty: they do not have electricity in their homes. However, the affordability element of energy security is not just a question of whether energy prices are low or high relative to disposable income. The *volatility* of prices is even more central. Price shocks often cause serious humanitarian or economic hardship, even political instability, as energy consumers struggle to cope with unexpected financial burdens.[9] Prices reflect market circumstances and signal market expectations, which in turn influence consumer choices and investment decisions, whether in favor of consumption or conservation. However, even in wealthy countries, when prices deviate seriously from established expectations, consumers find it hard to make rapid changes in their energy consumption.

The importance of price volatility has been evident in the United States in the last several years. The consumer who in 2001 bought a house in a far-flung U.S. suburb and a new sport utility vehicle (SUV) still needed to get to work in July 2008, when gasoline prices surpassed $4 per gallon. Only with some struggle would the consumer be able to get a new job that was closer to home, get a new home closer to work, trade in the SUV for a higher-efficiency vehicle, or start using public transit if it was available. On average, Americans spent only about 2 percent of their household income on gasoline in 2001; by the summer of 2008, gasoline accounted for 4.5 percent of household income.[10] In response, thousands of Americans who had the opportunity to drive less and use public transportation more frequently started to do so. In late 2008, the number of vehicle miles traveled (VMT) dropped 4.6 percent from the level in the previous year.[11] Public transit ridership grew to the highest level in twenty-five years, an increase of 6.5 percent between 2007 and 2008.[12] Americans also started changing their automobile buying habits, shifting to smaller and more efficient cars. Year-on-year sales of gas guzzlers dropped 2.6 percent in 2006, 10.5 percent in 2007, and more than 25 percent by early 2008.[13]

Consumers naturally tend to prefer inexpensive energy, at least in the short run, because low energy prices allow them to spend their disposable income on other things. The problem with energy policies that place a high priority on low prices is that low prices fail to convey the full impact of energy use. They are therefore incompatible with true energy security because the expectation of low prices encourages consumption, discourages investment in higher-efficiency manufacturing, discourages new energy development (especially for higher-cost, lower-emissions technologies), and makes buyers vulnerable to price shocks when their expectation proves wrong. One of the key unknowns on the minds of energy policymakers at this writing is whether the drop in fuel prices through the second half of 2008 will result in an early rebound of demand and thus an abrupt return to painful price increases.

National approaches to energy pricing both reflect and have a major influence on the political economy of countries. For example, while U.S. energy policies have traditionally emphasized inexpensive retail energy, since the 1970s western European and Japanese policies have sought to discourage consumption by using taxes to raise retail energy prices. European and Japanese energy pricing is enabled by the existence of good public transportation systems and land use policies that are conducive to mass transit, while transit and land use policies are dictated in part by high energy prices. On the U.S. side, however, the relative absence of public transportation means that almost every American voter feels pain when gasoline prices rise. At that stage, support for public transit increases, but land use policies that have encouraged suburban sprawl make it expensive to create transit systems that offer a cost-effective alternative to driving.

Getting prices right is one of the absolutely central prerequisites to enhancing energy security. Energy prices that convey the full cost of energy consumption stimulate appropriate consumer responses, but they also necessitate careful decisions at the level of individual households and enterprises. In the midst of a recession, it is easy to obstruct plans for even small, gradual increases in energy prices, and that reality is one of the key threats to current efforts to promote energy reform in the United States. But no time is an easy time to institute policies that convey the full cost of energy usage. For too long, Americans have allowed—and at times even encouraged—their political leaders to avoid the painful political debates needed to bring balance to the country's energy security. The nation persists in such policies at its own peril.

Sustainability

In the past, definitions of energy security typically did not include environmental considerations. However, a contemporary approach to energy security must place emphasis on environmental sustainability, for several reasons:

—*Energy infrastructure typically is long-lived.* Decisions made today have long-term implications for how energy is produced, converted, stored, and used. An automobile bought today will be used for at least three to five years, maybe longer. But even then, it will likely live on for a decade or longer in the hands of second-hand buyers. The coal-fired power plant that a utility company builds today will be an investment based on twenty-five years of use or longer; that means that decades of carbon emissions will stem from one near-term decision. With two new coal-fired plants coming online every week in China alone, current decisionmaking is creating the environmental reality that will shape people's lives around the world for decades to come.

—*Promoting energy security without including sustainability will promote use of technologies and practices that will exacerbate climate change.* For example, coal-to-liquids technology, for which Congress has debated possible subsidies designed to reduce petroleum use, would increase greenhouse gas emissions unless major breakthroughs occur in carbon capture and storage (CCS) to make it commercially and technologically viable. There is little chance of getting to any viable alternatives without pricing carbon to make CCS competitive with current technologies.

—*Climate change clearly will affect energy systems profoundly.* For example, rising sea levels will require redesign and re-construction of the transportation infrastructure that serves energy systems—from oil terminals to shoreline rail and road systems.

Some analysts and policymakers correctly note the current lack of the full suite of technological solutions needed to deliver the massive greenhouse gas emissions reductions that the scientific community now recommends.[14] These skeptics assert that because those solutions do not yet exist, it makes little sense to be concerned today with the integration of sustainability into current energy security calculations. That argument defies elemental logic. Given the magnitude of the changes that will be required throughout the global energy economy, given the enormity and longevity of the energy investments that are called for in the coming two decades, and given that individual consumers must participate actively in many of the marketplace changes that need to occur, the sooner nations reorient their thinking and buying, the better. Thus, talking about energy security without talking simultaneously about sustainability is, at best, penny wise and pound foolish.

Finally, it is worth noting one last issue regarding the definition of energy security: The broad term "energy security" risks obscuring the fact that contemporary economies employ energy in distinct forms for different purposes. The transportation sector relies heavily on petroleum in the form of gasoline, diesel fuel, jet fuel, and other petroleum-based products. Some substitution by other fuels is possible in the case of some automobiles, but almost all vehicles currently run on liquid fuels, most frequently petroleum-based fuels.[15] Because of the particular demands of the transportation sector, security of supply for oil and refined oil products is a key aspect of overall energy security for the United States.

Country-Specific Priorities

If the concept of energy security incorporates the four elements suggested above—availability, reliability, affordability, and sustainability of energy services—it is worthwhile to consider whether all countries evaluate their energy security priorities and vulnerabilities identically. Not surprisingly, they do not. That in no way detracts from the fundamental usefulness of the concept, however. Each country naturally faces a distinctive energy security position, and each country's policy priorities should reflect its uniqueness. One country's position in relation to the availability and affordability of energy services may be favorable, but it may face challenges in relation to reliability and sustainability. Within a given country, even individual regions and socioeconomic groups may have different positions because their location or economic condition means that they either do or do not have sufficient energy services at their disposal.

For example, China has traditionally viewed its chief energy security challenge as the need for reliable, uninterrupted supplies in order to feed a growing economy and maintain social stability. Consequently, reliance on domestically produced coal remains an important priority in spite of its obvious environmental impacts. With respect to crude oil, China's concerns about availability and reliability grew significantly greater as the country became a significant oil importer over the last decade. Most of China's oil imports transit the narrow Strait of Malacca. Today, more than 12 million barrels of oil transit the strait each day—to all destinations, not only to China. By 2030, due to the bullish growth in oil demand in China and across the rest of East Asia, oil transits through the strait are projected to double, creating a significant supply vulnerability in case of an accident or a terrorist attack.[16]

In light of those dynamics, it is understandable that policymakers in Beijing have been willing to consider what would in the past have been viewed as economically dubious investments, such as mammoth pipelines running from Xinjiang in the west of the country to the eastern industrial centers or, even more ambitiously, from Central Asia to Xinjiang and then eastward. Economic attractiveness (or affordability, as defined above) has simply been a lower priority for Beijing than availability and reliability.[17]

Other countries have made the calculated decision to sacrifice lower prices for greater reliability and (arguably) greater sustainability. For example, France, Japan, and Finland all rely heavily on nuclear power and have persisted in that policy despite the significant challenges that nuclear power faces in terms of costs and social acceptance.[18] The United States, by contrast, has to date emphasized relatively heavy use of plentiful and inexpensive domestic coal for power generation, despite the local and global environmental shortcomings of coal. In short, each country makes its own choices about how to optimize its own energy security position.

Even some major energy producer countries struggle to ensure that their people have available, reliable, affordable, and (though to a lesser degree) sustainable energy services in the proper places across their territories. For example, Iran, which has the world's second-largest proven reserves of natural gas and the third-largest reserves of oil, actually faces significant challenges with respect to energy security on the regional level within its borders. Iran's hydrocarbon deposits are clustered onshore and offshore along the coast of the Persian Gulf, while much of its energy demand is in the north, in and around Tehran. Iran's capital city would gladly consume more domestically produced gas, but much of that gas is used for reinjection into oil wells to prop up production of oil, the country's critical export commodity.

Many energy-rich countries and developing countries use artificially low and slow-rising consumer energy prices to compensate for other economic weaknesses. However, when consumers are protected from the full price of energy, "excessive" demand is created—more demand than would exist if prices reflected current market conditions. As prices spiked in the summer of 2008, this issue came to the fore in international discussions of turbulent oil markets. For example, in June 2008, Japan convened in Aomori City a meeting of energy ministers from the G-8 countries plus China, India, and the Republic of Korea, together with the head of the International Energy Agency (IEA). After the meeting, U.S. energy secretary Samuel Bodman commented testily to the press: "A lot of nations are still subsidizing oil, which ought to stop in our view. I don't have any belief the Chinese are going to [eliminate subsidies], but I repeat myself.

Governments should cease subsidizing the use of oil. Consumers are not paying high prices . . . and are not changing their habits."[19] In fact, in the weeks leading up to the G-8 meeting in Hokkaido, a range of developing countries, including India, Indonesia, Malaysia, Taiwan, and China, all raised their consumer energy prices.

Russia, the country with the world's largest natural gas reserves and significant oil reserves, faces significant domestic energy security challenges despite its petroleum wealth, its huge hydropower dams, and its highly developed nuclear industry. (Russia's actions affecting *other* countries' energy security are addressed in the accompanying text box.) Since the early 1990s, experts within Russia's electricity industry have warned political leaders about the massive quantity of investment required to refurbish and replace aging power generation, transmission, and distribution assets. Anatoliy Chubais, who served as the CEO of RAO United Energy Systems (UES) of Russia from 1998 until the company's dissolution in 2008, doggedly pursued higher tariffs and industry restructuring. Chubais stated again and again that the tariffs and restructuring were necessary to attract reinvestment. Nonetheless, the pace of new investment in Russia's power sector lagged.

As a consequence, Russia experienced a series of power blackouts, including in the capital. Interestingly, the logic of Chubais's case eventually made an impact on top-level Russian decisionmakers. At the very time when President Vladimir Putin was recentralizing decisionmaking in the Russian oil and gas industry, he approved RAO UES's restructuring and privatization program. It will be interesting to see whether the program will succeed in ending Russia's electricity supply security problems.

Priorities for Energy Security

In the spring of 2001, Vice President Cheney famously said, "Conservation may be a sign of personal virtue, but it is not a sufficient basis for a sound, comprehensive energy policy."[20] Instead of emphasizing conservation or efficiency, Cheney's national energy policy focused heavily on energy production, including the development of the Arctic National Wildlife Refuge. Given the immensity of U.S. energy consumption, that was a remarkable choice. The United States represents roughly 4 percent of the global population but accounts for more than 21 percent of total primary energy demand.[21]

Russia, Ukraine, and European Energy Security

Russia, in addition to facing energy security challenges within its own borders (see main text), has played a controversial role regarding the energy security of its neighbors and energy trading partners. Russia's role first gained broad attention in 2006. Just as Moscow was precipitating one of the most severe European energy security crises on record, Russian leaders announced that the Saint Petersburg G-8 summit would have energy security as one of its key topics. In early 2009 Europe received a stark reminder of Russia's willingness to flex its energy muscle: Gazprom cut off all gas flowing to and through Ukraine for a period of two weeks.

Uneasy Business Partners. Russia currently relies on Ukraine as its partner for critical exports of natural gas to buyers in OECD Europe. Ukraine's location and its vast Soviet-era gas transit infrastructure make it an essential link in the Eurasian gas value chain. More than 110 billion cubic meters (bcm) of natural gas flows across Ukraine to Europe every year. While it represents less than one-quarter of Russia's total gas production, it provides roughly two-thirds of Gazprom's revenue and therefore accounts for a major share of Russia's export earnings and government revenues. It also represents more than 20 percent of total gas consumption in the European Union. The stakes are high for all concerned.

When Viktor Yushchenko took over the Ukrainian presidency in early 2005, after the so-called Orange Revolution, he declared a new day in reform and policy-making. He said that his country would pursue a westward-leaning policy emphasizing economic reforms, EU accession, and membership in NATO—exactly the kind of stance that had led to the Kremlin's ham-handed support for his electoral opponent. In the spring of 2005, Yushchenko declared Ukraine's desire to conduct its gas relations with Russia on an all-cash basis and promised to bring to an end the highly opaque barter exchanges that had been the source of major corruption and bitter conflicts with Russia since 1992, during the administrations of presidents Kuchma and Kravchuk. Finally, Yushchenko reiterated Ukraine's long-standing refusal to sell its international gas transit pipelines to Gazprom.

Gas Crisis of 2006. Displeased by Yushchenko's policy priorities, Russia responded by declaring that it would abrogate its supply obligation under the existing gas agreement with Ukraine. Throughout the second half of 2005, Moscow and Kyiv engaged in an escalating war of words, trading accusations in the press rather than negotiating a new gas deal. The Ukrainian side appeared to presume that it could win a game of brinksmanship without endangering its own gas supply or the gas supply security of downstream neighbors in Europe. The Russian side appeared

(continued)

Russia, Ukraine, and European Energy Security (Continued)

to presume that if the confrontation reached a crisis, outside observers would blame Ukraine rather than Russia for the situation. Both were wrong.

On January 1, 2006, in the midst of some of the bitterest winter weather in decades, Russia cut gas supplies to Ukraine.[1] Immediately, the risks to European gas supplies were clear. Countries downstream from Ukraine reported reduced flows despite Russia's claim that only volumes destined for Ukraine had been cut. Officials in the United States and a number of European nations spoke out against Russia's action.[2] Russia quickly lost the war of global public opinion. In fact, Russia backed down on January 3 and restarted full gas shipments, the day before a new gas deal was reached.

The January 2006 gas agreement was a curious thing. It increased Ukraine's nominal price for imported gas to $95 per thousand cubic meters, fixed an artificially low transit fee for Russian gas across Ukraine to Europe, sanctified an expanded monopoly role inside Ukraine for a shady trading company called RosUkrEnergo, secured no commercial or legal protections for Ukraine, and consequently provided no commercial predictability for parties along the value chain. The logic of the deal was difficult for even a knowledgeable observer to fathom. Many concluded that the deal reflected the corrupt interests of well-positioned individuals rather than the interests of Ukraine, Russia, or Europe. Nonetheless, European institutions and governments breathed a big sigh of relief without looking closely enough to ascertain whether it was a durable solution to a conflict that

1. "Cutting gas supplies to Ukraine" is less straightforward than it may appear to the lay reader. As mentioned above, Russia relies on Ukraine to reach key consumers in OECD Europe. In the midst of winter, cutting off Ukraine entirely would be impossible unless Russia were prepared to cut off Europe simultaneously (which happened in 2009). If Russia took that course of action, it would face major difficulties because it would have nowhere to put the gas coming out of Gazprom's many wells. In January 2009, when Russia carried out a complete cut-off, Russia was forced to "shut in" its production wells, which caused hundreds of millions of dollars of monetary losses and risked damage to the underground reservoirs. In 2006, Russia therefore only reduced pressure on the gas transit lines by an amount that it said would be equivalent to the Ukrainian domestic share of the total volume. For more on the Ukrainian-Russian gas trade, see Jim Nichol and Steven Woehrel, "Russia's Cut-Off of Natural Gas to Ukraine: Context and Implications," RS 22378 (Congressional Research Service, February 15, 2005); see also Yulia Mostova, "Analysis and Commentary," *Zerkalo Nedeli (ZN) on the Web* 2 (581) January 21, 2006, and "It's a Gas—Funny Business in the Turkmen-Russian Gas Trade," *Global Witness*, April 2006. For a good chronology of the broader January 2009 cut-off, see Simon Pirani, Jonathan Stern, and Katja Yafimava, "The Russo-Ukrainian Gas Dispute of January 2009: A Comprehensive Assessment," Oxford Institute for Energy Studies, 2009 (www.oxfordenergy.org/pdfs/NG27.pdf).

2. For the U.S. statement see "Ukraine: Suspension of Gas Shipments from Russia January 1," statement to the media by U.S. State Department spokesman Sean McCormack, January 1, 2006 (www.state.gov/r/pa/prs/ps/2006/58613.htm). For coverage of European statements, see Andrew E. Kramer, "Russia Restores Most of Gas Flow Despite Dispute with Ukraine," *New York Times*, January 2, 2006 (www.nytimes.com/2006/01/02/international/europe/02cnd-gas.html).

Russia, Ukraine, and European Energy Security (Continued)

endangers gas supply security for Europe. Ideas about domestic Ukrainian gas sector reform went nowhere, despite the ambitious declarations after the Orange Revolution.

Gas Crisis of 2009. In January 2009, it became painfully clear that the January 2006 crisis was not the last. Through the latter half of 2008, just as in 2005, the parties failed to reach agreement on the sales-purchase price. That failure seemed surprising given the fact that in October 2008 the prime ministers of Russia and Ukraine had issued a memorandum of understanding that appeared to pave the way to more durable, more widely accepted provisions for gas sales-purchase and transit.

At the end of December 2008, negotiations broke down amid a hail of acrimonious charges. The Russian side accused the Ukrainian president and prime minister of being so internally divided that they were unable to reach agreement with each other on terms. Russian officials also claimed absurdly that the U.S. government was manipulating the situation to cause a crisis. Gazprom cut off all gas flowing to Ukraine and European consumers, and thousands of apartments in Ukraine, Romania, Bulgaria, and the Balkans grew stone cold. Officials of the European Union—after initially trying to avoid commenting on the matter—made clear that they did not care who was to blame for the crisis and that both sides needed simply to restore the flows and avoid humanitarian disaster.[3]

Finally, after nearly three weeks of crisis, the two sides signed a pair of agreements—one governing sales-purchases and the other transit of gas. The January 19 agreements signaled the end of the most recent gas crisis but not the ultimate resolution of the underlying issues. The new agreements included some of the features that stable, professional European gas deals include. But the new agreements also included several key provisions—and omitted several others—that mean instability will likely return before long.[4]

Sources of the Recurrent Russian-Ukrainian Crises. To understand the energy security implications of the Russian-Ukrainian gas conflicts, one needs to examine the many factors that, to a greater or lesser extent, contribute to the conflict. The first is the price of gas. Beginning in 2005, President Putin and many Russian commentators claimed that Ukraine had since 1991 received gas for artificially low prices, a subsidy that Russia stated it could no longer afford. If in

3. Pirani, Stern, and Yafimava, "The Russo-Ukrainian Gas Dispute of January 2009."
4. For a summary of key elements of the January 2009 gas deal, see Steven Pifer, Anders Aslund, and Jonathan Elkind, "Engaging Ukraine in 2009," Policy Paper 13 (Brookings, March 2009) (www.brookings.edu/~/media/Files/rc/papers/2009/03_ukraine_pifer/03_ukraine_pifer.pdf).

(continued)

Russia, Ukraine, and European Energy Security (Continued)

fact Russia was subsidizing Ukraine, it would only be natural for Russia to raise gas prices. However, that argument ignores the fact that for much of the period between 1991 and the end of 2008, Ukraine actually paid significantly higher prices for gas than one would conclude from the quoted, nominal prices.[5] In reality, after one normalizes for the cost of transit to different destinations, Ukraine's true price for gas imported from Russia was close to—and at times even higher than—the price paid by its central and western European neighbors.

A second factor that has clearly played into the recurring Russian-Ukrainian crises is Ukraine's failure to proceed with long-needed energy reform. Ukraine consumes prodigious amounts of natural gas—and produces far less from domestic reserves than it has the potential to produce—in large part because the economic fundamentals of the energy sector are subverted by current Ukrainian policy. Instead of proceeding with energy reform designed to promote transparency and economically rational outcomes, Ukraine has used the energy sector as a massive domestic subsidy vehicle. The regulated prices of natural gas for industrial, institutional, and residential consumers all fail to cover full costs. Inevitably, that means that Ukrainian taxpayers subsidize energy consumption, which, rather than promoting energy efficiency, only adds to demand.[6]

A third and final cause of the recurring Russian-Ukrainian gas crises has been Russia's exploitation of energy needs as an instrument of state power. In that regard, the Ukrainian situation is part of the much broader context of Russian energy policy, in which Russia not only struggles to ensure its own domestic energy security but also is perfectly happy to manipulate other countries' energy security vulnerabilities at the same time.[7]

5. The discrepancy between nominal and actual prices stemmed from the fact that Ukraine purchased its gas through a series of mysterious middleman companies—Itera, then EuralTransGaz, then RosUkrEnergo—that received payment in kind. For a discussion of Ukraine's nominal and real gas prices, see Edward Chow and Jonathan Elkind, "Where East Meets West: European Gas and Ukrainian Reality," *Washington Quarterly* 32, no. 1 (Washington: Center for Strategic and International Studies, January 2009).

6. For a fuller discussion of Ukrainian energy policy, see Chow and Elkind, "Where East Meets West."

7. In 1997 Russia and Turkmenistan failed to reach agreement on the price for gas purchased by Russia, and Russia cut off its purchase and shipment of Turkmen gas. Turkmenistan had no other gas export capacity and was forced to shut in wells. In 2000, on the eve of elections in Georgia, Gazprom cut gas supplies to Georgia. In January 2006, two gas pipelines and a high-voltage power line connecting Russia and Georgia mysteriously exploded one after another in one of the most heavily guarded parts of Russian territory. In January 2007, Russia cut shipments of crude oil along the Druzhba pipeline through Belarus, affecting refineries in Slovakia, Poland, and Germany. For more discussion of such cut-offs, see Keith Smith, "Russian Energy Policy and Its Challenge to Western Policy Makers," testimony before the House of Representatives Governmental Reform Subcommittee on Energy and Resources, May 16, 2006 (www.csis.org/media/csis/congress/ts060516smith.pdf).

Russia, Ukraine, and European Energy Security (Continued)

Russian officials have made no secret of their use of energy for political purposes. As global energy prices began what would ultimately be a steady five-year climb (beginning around the start of the Iraq war in early 2003 and peaking in the summer of 2008), senior Russian officials and commentators alike reveled in the Russian state's newfound influence and its status as a so-called energy superpower. President Putin himself, when visiting Ekaterinburg in October 2003 with German chancellor Gerhard Schroeder, snapped brusquely at reporters: "There should be no illusions. . . . In the gas sphere, they will deal with the state."[8] Meddlesome European Union officials seeking a speedy end to the alleged trade distortion of low-cost domestic gas supplies would simply have to recognize that fact.

A couple of years later, Kremlin spokesman Dmitry Peskov commented on the abrupt energy price hikes that Russia was demanding of its neighbors, linking energy issues to Russia's broader foreign policy objectives. If a neighbor wished to join NATO, it would be viewed as disloyal, and "if you are not loyal then you do it [make the jump to higher energy prices] immediately," he said.[9] For a time, Gazprom's own publicly released policy on the Commonwealth of Independent States explicitly acknowledged the political aspects of the company's pricing decisions.[10]

8. "Ot Redaktsii: Krepkiy tupik" [From the Editors: A Tough Dead-End], *Vedomosti*, October 10, 2003 (www.vedomosti.ru/stories/2003/10/10-47-01.html).
9. Stefan Wagstyl, "Kremlin Frets about Blame in Litvinenko Case," *Financial Times*, December 12, 2006.
10. Gazprom, press release, "Board of Dretors Reviews Gazprom's Pricing Policy for Former Soviet Union," December 19, 2006 (www.gazprom.ru/eng/news/2006/12/22031.shtml [July 16, 2008]).

Ever since the administration of President Jimmy Carter, political leaders on both sides of the aisle have shied away from telling Americans the obvious: to enhance its energy security, the nation needs to use energy much more efficiently and with less impact on the environment. The best way to achieve those outcomes in the short run is to use less energy. And to do that the United States needs to convey—through pricing, that blunt pocketbook reality hated by all—the full impacts of energy use on its economy, society, and environment.

In the course of the next two to three decades, developing and implementing sound energy policy will present a huge challenge to the United States. Truly comprehensive energy policy, to use Vice President Cheney's rubric, will require use of many different elements simultaneously, including supply-side measures, technology development, and commercialization efforts. But it is essential to focus on the uppermost policy priorities, without which the country simply will be unable to navigate the changes that it must make in its energy policy. The following three priorities must be central to U.S. energy policy in the coming period:

Priority 1: Energy efficiency must be a national pursuit. The United States needs to enshrine efficiency as its very top energy priority because doing so can help reduce its energy security vulnerability in a timely fashion while improving economic and environmental performance. Energy efficiency is an approach that has yielded proven results in the past; there is still great scope for improvements in efficiency; and seriousness about energy efficiency will improve the country's ability to undertake other cooperative energy security projects with international partners. A national energy efficiency program should include quantified efficiency targets that simultaneously support the goals of climate change policy; it should be comprehensive in coverage, touching energy supply as well as all end users, whether individuals or companies; it should be market based; and it should be complemented by expanded support for research and development on energy efficiency and renewable energy.

Energy efficiency has been one element of U.S. energy policy since the 1970s, and its inclusion has resulted in

—federal and state legislation on performance standards and other policy tools
—a base level of public awareness
—numerous policy innovations
—significant reductions in energy use below the amount that would otherwise have been used.

Unfortunately, energy efficiency has never been treated as a cornerstone of policy; much more typically, it has been an afterthought. For example, the Energy Policy Act (EPAct) of 2005, which was debated in Congress for four and a half years, buries modest energy efficiency provisions under a mound of other measures intended to promote cheap energy—provisions supporting coal, nuclear energy, oil and gas development, siting, and regulation. The law did not favor efficiency over new production.[22]

That was not accidental. When President Carter sat in front of the fireplace in his cardigan, two weeks after entering office, he told Americans that they needed to "sacrifice" in order to ease the energy crisis of the

1970s.[23] Carter said energy conservation would be central to his policies: "Our failure to . . . take conservation seriously started long before this winter and will take much longer to solve." Later in the same year, he declared that the effort to place conservation at the heart of a national energy policy was a matter of the utmost seriousness, "the moral equivalent of war."[24]

The problem is that Americans despised Carter's cardigan-wearing sobriety. Energy conservation was lampooned as "shivering in the dark"— a matter of turning off lights and enduring less comfortable homes and offices rather than sustaining comfort and quality of life through more thrifty design and better end-use technologies. Since those days, U.S. leaders—Republicans and Democrats, legislators and executive branch officials, labor leaders and industrialists alike—united around the pursuit of low energy prices and the avoidance of anything vaguely resembling "sacrifice." The emphasis on cheap energy supply has had the inevitable result that the United States consumes more energy than other industrialized countries that have placed greater priority on efficiency and conservation.

Historically, where energy efficiency has succeeded in the United States, it has been to a large degree thanks to technology standards. Automobiles, air conditioners, refrigerators, and other appliances sold on the U.S. market must carry efficiency rating labels, and often they must meet minimum performance standards. In that form, efficiency standards are a proven tool and have already saved the country both huge quantities of energy and great sums of money that would otherwise have been spent unproductively. According to the Bush administration's own National Energy Policy report of May 2001, "Had energy use kept pace with economic growth, the nation would have consumed 171 quadrillion British thermal units (Btus) last year instead of 99 quadrillion Btus. About a third to a half of these savings resulted from shifts in the economy. The other half to two-thirds resulted from greater energy efficiency."[25]

Yet energy efficiency is an area in which there are abundant opportunities for further cost-effective improvements across the United States. To quantify the potential for energy efficiency improvements is a difficult task, and estimates vary widely due to differing analytical approaches. Even so, many studies suggest that it would be economically realistic to reduce natural gas consumption by 9 percent or more and electricity consumption by 20 percent or more, with the right set of policies and measures. Some estimates range considerably higher.[26] The United Nations Foundation study

"Realizing the Potential of Energy Efficiency" recommends establishing a goal for G-8 countries of annual increases in energy efficiency of 2.5 percent a year, roughly twice the rate of improvement that currently occurs.[27]

One of the complications facing policymakers who wish to give priority to energy efficiency is that efficiency improvements result from decentralized decisions by myriad economic actors. Promoting efficiency thus requires an array of policy approaches and specific measures. Energy consumption levels also reflect choices of equipment and infrastructure that are purchased and then used for a long time (refrigerators, automobiles, heating and air conditioning systems). That is why early emphasis on energy efficiency is crucial.

To promote efficiency effectively, the United States needs to employ all the available tools, including pricing structures that will create the right incentives for the purchase of high-efficiency devices; programs to inform consumers about the energy requirements that their purchases will commit them to (such as the Energy Star program); legal and regulatory reform in specific instances; and in some cases, new or strengthened minimum performance standards for goods and equipment, including household and commercial appliances, building materials, and automobiles and light trucks through instruments like corporate average fuel economy (CAFE) standards. If the United States does employ a broad and aggressive approach on energy efficiency, there is every reason to believe that it could significantly reduce the energy intensity of the economy. According to the most recent data from the Energy Information Administration, the U.S. economy consumes 8,841 British thermal units for every dollar of gross domestic product (GDP). Japan consumes about 27 percent less, and European countries on average consume 26 percent less.[28]

In certain areas, policies and measures that have been employed to date still fail to address market failures or other structural challenges that obstruct broader adoption of energy efficiency standards. Often that is due to the so-called principal-agent problem, wherein the benefits or costs of an action accrue not to the person taking the action but to a different party. U.S. policymakers should spur new policy innovations in these areas and should make clear that they are prepared to consider innovative approaches to achieve energy efficiency improvements. Ideas abound. For example, home builders typically lack the incentive to install highly efficient appliances and materials in new homes because they raise the ultimate purchase price of the home, discouraging potential buyers. But if rate structures were amended, electric utility companies could be enticed to bear some of the incremental cost to build highly efficient homes because the utility companies could avoid more costly investments in new power plants.[29]

Another policy gap that could be filled through innovative measures relates to the shortcomings of existing CAFE standards. As automobiles grow stingier with fuel, the temptation for consumers is to drive more, because the cost of fuel for a consumer's new car does not hurt the house-hold budget quite as much as the cost of fuel for the old car. To avoid this unintended incentive to drive more, either a direct or indirect approach could be employed. The most direct and efficient approach would be to implement a tax on fuel, which would provide a completely transparent impetus to avoid driving. If that approach is not politically viable, an indirect instrument could be used as an alternative. Jason Bordoff (who writes on a different topic in chapter 9 of this volume) has suggested the introduction of "pay-as-you-drive" automobile insurance, which would give drivers an economic incentive to avoid increasing their vehicle miles traveled.[30]

If President Obama makes energy efficiency an explicit priority, a side benefit would be to improve the atmosphere for international cooperation on all issues related to energy security. As discussed in greater length below, international cooperation is an essential prerequisite for the effective han-dling of many energy security issues, not the least being climate change.

Priority 2: Decisive action must be taken on climate policy. Voters have had energy and climate on their minds as prices and global temperatures climbed over recent years, and their concern has been acknowledged—at least rhetorically—by political leaders. In his State of the Union address in 2006, President George W. Bush, a former Texas oilman, proclaimed that the United States was "addicted to oil." A year later, Bush announced the goal of reducing gasoline consumption by 20 percent below the levels currently projected (under a business-as-usual scenario) for the year 2010.[31] With a new Democratic majority, the 110th Congress started focusing on energy and climate as it began its work in January 2007. In June 2008, the closely divided U.S. Senate briefly debated a bill to insti-tute a cap-and-trade system for carbon dioxide and other greenhouse gases. Through early 2009, representatives Henry Waxman (D-Calif.) and Edward Markey (D-Mass.) were leading consideration of another cap-and-trade bill in the U.S. House of Representatives.

With all this attention to issues that long struggled to gain national prominence, one might be tempted to conclude that the United States is close to new legislation on climate change. Sadly, in my view, that conclusion may still be premature. It is true that for the first time both the president and the congressional leadership have declared the goal of enacting a cap-and-trade bill into law. It also is true that the Obama administration has sought to embed green energy investments in the economic stimulus package and to make the point that changing U.S. energy consumption patterns can be the engine of new economic development. However, in trying to guess how long enactment of a climate bill might take, one would do well to reflect on two points: First, it is worthwhile to recall the experience under the Bush administration, which entered office in 2001 claiming a mandate to pass the first comprehensive energy legislation since 1992. President Bush had Republican majorities in both houses of Congress, which were eager to support their leader in the White House. Nonetheless, the Energy Policy Act (EPAct) of 2005 was signed into law only four and half years later. The second consideration is the magnitude of the climate change issue and the profound ways in which any comprehensive climate legislation will, over time, affect daily life in the United States. Housing, transportation, jobs, leisure and recreation—all, over time, will need to reflect the fact that the U.S. population now lives in a carbon-constrained world. The issues to be resolved are complex, and every decision has the potential to engender opposition from one or another powerful lobby. In June 2008, the Senate debate on climate legislation focused almost entirely on the charge that it would result in higher energy costs at a time of record prices at the gasoline pump. In 2009, the recession is being used as justification for not instituting a climate change bill. How can one talk about raising energy prices (no matter how modestly or slowly) when U.S. families already are struggling?

If the impulse is to simplify matters by starting with a climate bill that is not comprehensive, that opens a different can of worms. The industries that are regulated under a given climate bill will protest that they are being treated in a discriminatory manner. Already one occasionally sees inter-industry strafing of that sort between the electricity industry and the oil and gas industry. The representatives of each make unfavorable comments about the perceived preferences being showered on the other.

Not the least obstacle to climate legislation is the fact that it will result in higher energy prices; in fact, that is exactly the point. At present, energy prices completely fail to place a price on the environmental and security costs—the "externalities"—of current energy use. The United States must establish a system for pricing the emission of carbon and other greenhouse gases, at prices that should start low and grow steadily over time. Nonetheless, in a time of energy price volatility—and deep recession—imposing even incremental additional costs for energy will be politically tricky. It is, however, essential. The country must begin the work that will last for the rest of our children's lives and beyond. It must give the marketplace a signal that the cost of carbon emissions is going to rise inexorably over the coming decades. Policymakers must work calmly and steadily toward the goal of adopting an approach that minimizes the risk of political reversals. Only that kind of signal will create adequate incentives for new technology development (including make-or-break technologies like carbon capture and storage for coal usage), low-carbon energy investment, efficient transportation infrastructure, and the other ingredients of a low-carbon future.

Priority 3: Energy policymakers must be required to take a "Hippocratic oath." Newly trained doctors are admonished, "Primum non nocere"—above all else, do no harm.[32] The same spirit needs to guide policymakers in the coming years to ensure that the United States truly improves its energy security, including climate protection. The nation needs to acknowledge the fact that the period that it is now entering presents gargantuan challenges for energy policy. It must undertake fundamental reform of technologies, practices, fuels, investment, and economics in the global energy sector, but it needs to do so without gratuitously destabilizing the current state of affairs—to do so at least with a minimum of disruption. After all, the energy sector is one of the most fundamental aspects of national and global economies. In other words, nearly everything should be changed, but in a way that keeps investment flowing, keeps voters happy (lest they vote out the scoundrels who bring them greener, unpopular energy policies), and keeps the lights on. That's a tall order, to be sure.

Think about the complexity of the task. For example, if policymakers think about energy security only in a narrow, outdated sense—as anything that leaves the country less dependent on imported energy supplies—they will be tempted to resort to the wholesale use of domestic coal reserves, even before the resulting carbon emissions can be successfully captured and stored. For transportation, they will allow aggressive tapping of Canadian tar sands (a "foreign" supply, yes, but significantly less risky, according to most commentators), which yields crude oil only after energy- and water-intensive processing and broad disturbance of the natural environment. They might also permit development of—worse yet, subsidize— new liquid fuels derived from carbon-heavy coal.

On the other hand, if policymakers act rashly and do not analyze with care the economic and social impacts of their climate protection efforts, there will be a great risk of policy reversals. If people find the climate measures too intrusive, too expensive, or just too annoying, then they will wait for the first available opportunity to repeal them. For the purpose of issuing a clear signal to the marketplace, that would be the worst of all worlds. On-again, off-again climate policy will chill investment in climate-friendly energy technology just when the country needs to kick off its wholesale program to improve comprehensive energy security.

The World Resources Institute has published a graphic (reproduced here as figure 6-1) that illustrates in a rough manner the potential interaction between traditional, outdated energy security concepts and the climate change characteristics of different technologies. True energy security will come as technologies listed in the top-right quadrant of figure 6-1 are employed—those that emit as little as possible of the greenhouse gases but that do not create problems with respect to availability, reliability, and affordability.[33] To do that, policymakers need to act carefully and make the outlines of future energy policymaking available to the market.

Conclusion

In the United States and around the world, the need to pursue a new vision of energy security—one based on availability, reliability, affordability, *and* sustainability—poses enormous challenges. It can be said with justification that success in this endeavor will be an existential matter. Within the global context, the United States enjoys great opportunities, but it bears special responsibilities as well. For a long time, the nation has thought about energy as a cheap input for its economy and lifestyle. Consequently, it has grown into an energy-intensive economy, and it has had to endure periods of great tension when its energy insecurity came directly to its attention in the form of an embargo, a hurricane, or a price spike. Now the country needs to look with greater care at how it wastes energy and at opportunities to significantly reduce its energy consumption. Policy reform must be incremental but unmistakable and concerted, focusing simultaneously on sustaining investment and promoting innovation. The nation's long-delayed comprehensive response to climate change must begin. Throughout that effort, policymakers need to bear in mind the essential importance of playing a short and long game simultaneously.

F I G U R E 6 - 1 . A Snapshot of Selected U.S. Energy Options Today

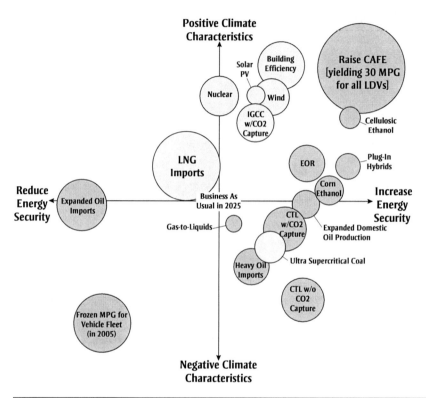

A Snaphot of Selected U.S. Energy Options Today:
Climate Change and Energy Security Impacts and Trade-offs in 2025

This chart compares the energy security and climate characteristics of different energy options. Bubble size corresponds to energy provided or avoided in 2025. The reference point is the "business as usual" mix in 2025. The horizontal axis (energy security) includes sustainability as well as traditional aspects of sufficiency, reliability, and affordability. The vertical axis (climate characteristics) illustrates lifecycle greenhouse gas intensity. Bubble placements are based on quantitative analysis and WRI expert judgment.

○ Power Sector (this size corresponds to 20 billion kWh)
○ Transport Sector (this size corresponds to 100 thousand barrels of oil per day)

For specific details on the assumptions underlying the options on this chart, go to www.wri.org/usenergyoptions
Revised: 7/2/2007

In conclusion, the United States needs a unique combination: a vision of a very different energy future, the discipline to pursue that vision in a way that fosters technology and encourages investment, and the patience to realize that it will need to persist in the process through times of progress and times of challenge.

Notes

1. See the first two Bush-Gore presidential debates, from October 3 and October 11, 2000 (www.debates.org/pages/his_2000.html).

2. For example, see Daniel Yergin, "Ensuring Energy Security," *Foreign Affairs*, March–April 2006.

3. For example, see Stephanie Seay, "Canadian Hydrate Development 20 Years Off," *Platts Oilgram News* 86, no. 133 (July 8, 2008), p. 9.

4. See Russell Gold, Bhushan Bahree, and Thaddeus Herrick, "Storm Leaves Gulf Coast Devastated," *Wall Street Journal*, August 31, 2005. Also see "Post-Katrina U.S. Gulf Shut-Ins at 870,374 Barrels/Day, or 58.02 Percent: Minerals Management Service," *Platt's Global Alert*, September 6, 2005. Also see David Ivanovich, "Southern Exposure Reveals a Weakness," *Houston Chronicle*, September 4, 2005.

5. R. James Woolsey, "High Cost of Crude: The New Currency of Foreign Policy," testimony before the U.S. Senate Committee on Foreign Relations, November 16, 2005 (http://foreign.senate.gov/testimony/2005/WoolseyTestimony051116.pdf).

6. One of the perennial temptations is to draw on the SPR during events that do not constitute a true market crisis. The most recent (though certainly not the only) example of that temptation occurred in July 2008, when some members of Congress called on President Bush to draw on the SPR to help alleviate high consumer gasoline prices. Politically attractive though it may be, to use the SPR merely as a price modulation device would be unwise and could undercut the reserve's availability in times of true crisis.

7. Obviously, not all energy systems are truly global. In fact, one of the benefits of renewable power sources like wind or solar energy is that they are not at the mercy of global energy markets.

8. For more on the importance of information for reliable energy systems, see Yergin, "Ensuring Energy Security."

9. For a poignant discussion of the effects of price volatility in impoverished countries, see Chip Cummins, "As Fuel Prices Soar, A Country Unravels: Energy Shock Hits the Upwardly Mobile Poor Hardest in Africa's Guinea," *Wall Street Journal*, November 18, 2006, p. A1.

10. Roger Lowenstein, "What Price Oil?" *New York Times*, October 19, 2008.

11. American Public Transportation Association, "Public Transit Ridership Surges as Gas Prices Decline," press release, December 8, 2008 (www.apta.com/media/releases/081208_ridership_surges.cfm).

12. The cited increase is a comparison of data for the third quarter of 2007 and the third quarter of 2008 and is especially noteworthy in that it is consistent with increases that were occurring as gasoline prices were still rising, which they ceased to do in July

2008. See American Public Transportation Association, "Public Transit Ridership Surges as Gas Prices Decline."

13. Ana Campoy, "Americans Start to Curb Their Thirst for Gasoline," *Wall Street Journal*, March 3, 2008.

14. The Intergovernmental Panel on Climate Change has recommended limiting atmospheric concentrations of greenhouse gases to 450–550 parts per million of CO_2 equivalent (CO_2e). That translates into the requirement to return global greenhouse gas emissions to roughly 50 percent of 1990 levels by 2050 and stabilize them at that level. See the IPCC's Fourth Assessment Report (www.ipcc.ch/ipccreports/ar4-syr.htm).

15. This reality is highlighted in David Sandalow, *Freedom From Oil* (New York: McGraw Hill, 2008).

16. See Kensuke Kanekiyo, Institute of Energy Economics, Japan, "The Asian Energy Outlook and Regional Cooperation," a presentation at the Japan–United States–China Trilateral Conference, Washington, November 10–11, 2006.

17. That is not to suggest that Chinese authorities pay no attention to sustainability. Even as Beijing continues its headlong building program for new coal-fired power plants, it acknowledges the local and global risks associated with power plant emissions. Beijing also is instituting ambitious new policies to promote automobile efficiency, among other provisions.

18. In most countries, nuclear power remains a source of great controversy. It does not result in the emission of carbon dioxide or other greenhouse gases, and that is clearly its strong suit in a carbon-constrained world. But the capital costs of building new nuclear power capacity are staggering, and the environmental challenges of fuel production and waste disposal are considerable.

19. Daniel Goldstein and others, "International Energy Agency Leader Sees Subsidy Cuts by Asian Nations Denting Demand," *Platts Oilgram News* 86, no. 113 (June 10, 2008), p. 11.

20. Joseph Kahn, "Cheney Promotes Increasing Supply as Energy Policy," *New York Times*, May 1, 2001.

21. *BP Statistical Review of World Energy 2008* (www.bp.com/liveassets/bp_internet/globalbp/globalbp_uk_english/reports_and_publications/statistical_energy_review_2008/STAGING/local_assets/downloads/pdf/primary_table_of_primary_energy_consumption_2008.pdf).

22. For an analysis of the provisions of EPAct 2005, see "Energy Policy Act of 2005: Summary and Analysis of Enacted Provisions," Congressional Research Service, March 8, 2006, Order code RL 33302 (www.bp.com/liveassets/bp_internet/globalbp/globalbp_uk_english/reports_and_publications/statistical_energy_review_2008/STAGING/local_assets/downloads/pdf/primary_table_of_primary_energy_consumption_ 2008.pdf).

23. President Jimmy Carter, "Report to the American People on Energy," February 2, 1977 (http://millercenter.org/scripps/archive/speeches/detail/3396).

24. President Jimmy Carter, "Address to the Nation on Energy," April 18, 1977 (http://millercenter.org/scripps/archive/speeches/detail/3398).

25. The White House, "National Energy Policy," report of the National Energy Policy Development Group, May 2001.

26. One of the complications of such estimates is the difference between the *technical* potential (the quantity of reductions that could theoretically be achieved through

the application of available technology, without regard to cost-effectiveness), *economic* potential (those reductions that would be cost-effective), and *achievable* potential (reductions that would actually occur given current policy, regulation, and economic factors). Another complication is secondary economic effects. For example, corporate average fuel economy standards made automobiles more efficient through the 1970s and 1980s, thus reducing consumers' out-of-pocket costs for driving a given distance. That cost reduction spurred unforeseen increases in the amount that the average driver traveled by car over the same period. For more on estimates of efficiency potential in the U.S. economy, see Steven Nadel, Anna Shipley, and R. Neal Elliott, "The Technical, Economic, and Achievable Potential for Energy Efficiency in the U.S.: A Meta-Analysis of Recent Studies" (American Council for an Energy-Efficient Economy, 2004).

27. The UN Foundation further recommends that this goal be agreed on by all the members of the G-8, which would also be an important objective. See United Nations Foundation, "Realizing the Potential of Energy Efficiency: Targets, Policies, and Measures for G-8 Countries" (Washington: United Nations Foundation, July 2007), p. 7.

28. U.S. Energy Information Administration, *International Energy Outlook 2006* (www.eia.doe.gov/pub/international/iealf/table1p.xls).

29. For other creative ideas about improving energy efficiency, see UN Foundation, "Realizing the Potential of Energy Efficiency." The example of utility investments in efficient homes is drawn from Timothy E. Wirth, Vinod Khosla, and John D. Podesta, "Change the Rules, Change the Future," May 22, 2007 (www.grist.org/comments/soapbox/2007/05/22/change/index.html).

30. Jason Bordoff, "Pay-As-You-Drive Car Insurance," *Democracy: A Journal of Ideas* 8 (Spring 2008) (www.democracyjournal.org/article.php?ID=6610).

31. Notably, President Bush's initiative was not a 20 percent reduction in current consumption levels, but a reduction below projected levels (a higher baseline). The target was to be met by a combination of higher fuel efficiency and "alternative" fuels—partially from ethanol and partially from coal-to-liquid and other nontraditional fuels. See the White House, "Twenty in Ten: Strengthening America's Energy Security," fact sheet, January 23, 2007.

32. The injunction *primum non nocere* is widely referred to as the Hippocratic oath, although the reference is somewhat erroneous.

33. World Resources Institute, "Climate Change and Energy Security Impacts and Tradeoffs in 2025," May 11, 2007 (www.wri.org/climate/topic_content.cfm?cid=4368).

CHAPTER SEVEN

Global Governance and Energy

ANN FLORINI

O ver the past several years, energy policy has assumed a prominent role on national policy agendas around the world. Yet there has been remarkably little effective coordination across borders on energy issues. In the absence of such coordination, it is unlikely that any national government will be able to develop and sustain energy policies that can balance the competing objectives of affordable energy services, reliable supply, environmental sustainability, and geopolitical security.

There is no single overarching international organization that is mandated to address any one of the collective action issues that energy policy poses, and that is not accidental. Rather, it reflects the way that the global energy sector has evolved over time. For instance, the International Energy Agency (IEA), despite its name, is actually a creature of the Organization of Economic Cooperation and Development (OECD); it has only twenty-seven members, and it addresses only a small portion of the energy issues

An earlier version of this chapter was presented at the Stanford Workshop on Managing Global Insecurities, March 16–17, 2007. It is available as Working Paper 1 of the Centre on Asia and Globalisation, Lee Kuan Yew School of Public Policy, National University of Singapore (www.lkyspp.nus.edu.sg/CAG). The author wishes to thank Saleena Saleem for excellent research assistance. The author is grateful to the Singaporean Ministry of Education for Grant T208A4109, which has supported elements of the work reported here. Any opinions, findings, and conclusions or recommendations expressed in this chapter are those of the author and do not necessarily reflect the views of the Singaporean Ministry of Education.

outlined in this chapter. Furthermore, the IEA has no regulatory powers and is ill-equipped to play any oversight role.

Energy needs are met largely through market forces (often heavily distorted by government policies), but global energy markets are extremely volatile and poorly regulated. The governments that regularly distort energy policies would also make up the bodies that would have to foster a mandate for an energy-governing entity. Thus, it is unlikely that a single international organization would emerge to address energy issues. Instead, progress is more likely to occur through relatively incremental changes in the mandates and performance of the multitude of relevant institutions. Taken together, those changes could facilitate a significantly improved global environment for good energy policy.

This chapter presents a preliminary examination of the formal institutions of global governance that have been tasked with addressing various components of energy policy. These formal intergovernmental bodies are only a piece of the energy governance puzzle, but they are crucial. Nonetheless, they have largely escaped systematic analytic scrutiny with regard to their impact on energy policy.

The chapter begins by identifying the broad range of problems that international energy policy should address and then looks at the existing mechanisms for addressing those problems. It continues with a brief history of how and why some of those institutions arose and an analysis of how well they deal with the global energy agenda; the chapter concludes with an assessment of various options for improving global energy governance.

Where Governance Is Needed

It is clear that global energy policy is in urgent need of dramatic change. According to a 2008 International Energy Agency report, staying on our present path would bring about a 70 percent increase in oil demand by 2050 and a 130 percent rise in CO_2 emissions. A rise of such magnitude would have a significant, perhaps disastrous, impact on the environment.[1] The IEA report, which presented various scenarios on deep emission reductions, stated that in order to bring the global CO_2 emissions back to current levels by 2050, an estimated $17 trillion in additional investments would be needed. That estimate assumes that the technologies already exist or are in an advanced state of development. If CO_2 emissions were reduced by 50 percent from the current levels by 2050, the more ambitious goal that climate scientists argue must be achieved to stave off potentially catastrophic climate change, the additional investments needed would be a massive $45 trillion.[2] But money is only part of the problem. Changing course to a politically and environmentally sustainable energy system for the world would require even more—not in terms of funding, but in terms of institutional and organizational development, along with a hefty helping of political leadership.

The first step in evaluating the state of energy governance is to define what problems governance is needed to solve. Managing the supply of and demand for energy involves dealing with four issues that require cooperation across borders:

—energy security
—environmental sustainability
—economic development
—respect for human rights.[3]

Energy Security

Energy security—that is, reliable and affordable access to energy supplies— is inextricably tied up with military and national security. Ever since the British converted their fleet from coal to oil on the eve of World War I to make it faster than its German counterpart,[4] major powers have looked on access to oil as a vital national interest, and any threats to that access may trigger a military response. The attack on Pearl Harbor—triggered when the United States, which supplied the vast majority of Japan's oil, responded to Japan's invasion of Indochina by freezing Japan's U.S. assets and cutting off oil exports—has been described as the "first energy war."[5] After the 1973 oil price shock, Henry Kissinger argued that U.S. security had been directly affected:

> In the last three decades, we have become so increasingly dependent on imported energy that today our economy and well-being are hostage to decisions made by nations thousand of miles away. . . . The energy crisis has placed at risk all of this nation's objectives in the world. It has mortgaged our economy and made our foreign policy vulnerable to unprecedented pressures.[6]

Concerns about such vulnerabilities and fears that competition over energy resources could turn violent continue today. As one recent report noted, "with new oilfields being discovered at a slowing rate and alternative energy yet to fully deliver on its promise, the resulting competition, and attempts to secure their safe delivery, could constitute a potential trigger for interstate tensions, even conflict."[7]

Oil vulnerabilities and tensions are, however, only a portion of the problem. Electricity shortages and blackouts have disrupted life in the United States, Europe, Russia, and many developing countries. As the market for natural gas expands both regionally and globally, new vulnerabilities emerge in that sector. Al Qaeda has threatened to attack the world's critical economic infrastructure, of which energy clearly is a key component. As leading energy analyst Daniel Yergin has pointed out, the challenges of energy security are enormous and growing:

> In the United States alone, there are more than 150 refineries, 4,000 offshore platforms, 160,000 miles of oil pipelines, facilities to handle 15 million barrels of oil a day of imports and exports, 10,400 power plants, 160,000 miles of high-voltage electric power transmission lines, and 1.4 million miles of natural gas pipelines. None of the world's complex, integrated supply chains were built with security, defined in this broad way, in mind. . . . The challenge of energy security will grow more urgent in the years ahead, because the scale of the global trade in energy will grow substantially as world markets become more integrated. Currently, every day some 40 million barrels of oil cross oceans on tankers; by 2020 that number could jump to 67 million. . . . The amount of natural gas crossing oceans as LNG [liquefied natural gas] will triple to 460 million tons by 2020. . . . Assuring the security of global energy markets will require coordination on both an international and a national basis among companies and governments, including energy, environmental, military, law enforcement, and intelligence agencies.[8]

The public debate often confuses energy *security* with national energy *independence,* which would require a country to meet its energy needs from sources within its own borders. Energy independence is neither feasible for most countries nor especially desirable as a goal in itself. Dependence on a world market that functions well is beneficial, not harmful—and that is as true for energy as for all other globally traded goods and services, for which specialization and trade demonstrably lower costs and increase economic efficiency for all.[9] Energy policy needs to ensure that markets function reliably and efficiently—a classic governance task.

It is not market forces per se that are the problem, but vulnerability to supply disruption and price shocks. Such problems occur in part because energy sources, especially oil, are not evenly distributed around the world. Instead, a large and rising share of the world's known oil reserves are concentrated in a handful of largely volatile and unstable countries—notably in the Middle East, Russia, Nigeria, and Venezuela. Moreover, energy markets, especially the oil market, suffer from significant market distortions, given that most oil supplies are controlled by a handful of government-dominated firms.[10] Energy markets in virtually all countries suffer from varying degrees of distortion by subsidies and taxes. Although some distortions, such as high European petrol taxes, are aimed primarily at addressing a public goods problem, many provide economic rents to powerful sectors, making it politically difficult to change policies to bring about more economically rational energy markets.

For example, energy interests often can be served by shifting regulatory policies from the state to the federal level through rent seeking. A classic example would be the Eastern coal industry convincing the federal government to amend the 1977 Clean Air Act. The amendment required all coal producers to reach a percentage reduction in emissions, requiring the installation of "scrubbers" at new coal-producing facilities regardless of the sulphur content of the coal. Before, the Western coal industry, which produced low-sulphur coal, had a competitive cost advantage over the Eastern coal industry, which had produced high-sulphur coal. Since all facilities now had to install "scrubbers," coal purchasers had less of a price incentive to transport Western clean coal across the country. The act effectively leveled the playing field by removing the low-sulphur cost advantage of the Western coal industry.[11] More recently, the automobile industry in the United States successfully lobbied the federal government for less stringent emissions standards after state law in California required progressively lower-emitting vehicles in the 1990s. The result was that the other states were free to adopt the less stringent federal emissions standard or California's more stringent emissions standard, but not both.[12]

Environmental Sustainability

To date, the major negative environmental externalities associated with energy production have been those associated with the extraction and consumption of fossil fuels. Such fuels constitute the overwhelming share of primary energy sources. As figure 7-1 shows, the situation is unlikely

FIGURE 7-1. Energy Demand, by Fuel

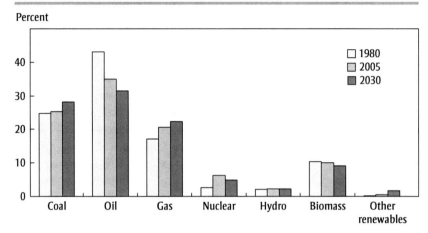

Source: Modified from IEA, *World Energy Outlook 2007*, p. 74.

to change for some decades. The environmental consequences of continued fossil fuel dependence are severe. By far the lion's share of attention is currently focused on climate change. Because climate change is the subject of other analyses in this book, it is not addressed in detail here, although this chapter assumes the necessity of a carbon-constrained future. (The latest Intergovernmental Panel on Climate Change report makes clear the overwhelming scientific consensus that human activities are responsible for an unnatural accumulation of greenhouse gases that are causing the atmosphere and oceans to become warmer.)[13]

The burning of fossil fuels also creates other major problems, such as smog and acid rain, issues that have come to plague some emerging-market countries at alarming levels. Transportation of oil leads to contamination of the marine environment, most dramatically in the form of oil spills but also through normal operation of offshore wells, washing out of oil tankers, and storage tank leaks, in addition to spill-off from land-based activities.[14]

Yet most nonfossil fuel energy sources pose their own environmental challenges. Nuclear energy technology at present involves such highly proliferation-prone and toxic materials as uranium and plutonium, some isotopes of which need to be safely stored for many thousands of years. The excitement in recent years over biofuels threatens to overlook some unpleasant realities that have begun to dampen expectations for their role in an environmentally sustainable energy policy. Although these fuels are carbon neutral (they absorb carbon dioxide while growing, then release it when burned), cultivation of crops for fuel raises serious environmental and social dilemmas—soil degradation, deforestation (with associated greenhouse gas emissions when, as is often the case, land is cleared by burning), and "food or fuel" competition over the best use of crops. Furthermore, the potential for biofuels to displace fossil fuels is limited because there is just not enough land or water to produce the quantities needed without negative environmental impacts.[15]

Other alternative energy sources also raise environmental issues, although generally to a lesser extent. Hydropower, a major generator of electricity in many countries, requires the construction of large dams, which frequently wreak local environmental havoc and can displace thousands or millions of people.[16] Solar photovoltaic (PV) cells contain toxic substances, and their energy must be stored in batteries, which also contain toxins. Wind power works only in certain locations and is generated by huge metal turbines whose manufacture requires substantial energy. Furthermore, several types of environmental damage are associated with the transmission systems that deliver renewable power to points of consumption: land use conflicts, soil erosion, destruction of forests and natural habitat, noise and interference with radio and television, deleterious effects on birds that collide with power lines and towers, electrocutions, the use of chemical herbicides and vegetation management techniques along rights-of-way, and the human health effects of exposure to electric and magnetic fields (which may contribute to childhood cancer).[17]

In short, technology to date has failed to produce any truly environmentally benign large-scale means to satisfy humanity's apparently insatiable appetite for energy. The need to shift from carbon-intensive energy sources is undeniable, but the alternatives need careful consideration.

Development

While growing use of modern energy sources, particularly fossil fuels, is creating one set of major governance challenges, the lack of access to modern energy sources is creating another. Current energy policies have failed to address the needs of vast numbers of people. Nearly 2 billion lack access to electricity, which is essential to a decent quality of life. The IEA has noted that the number of people who use dirty traditional biomass for cooking will grow from 2.5 billion to 2.7 billion by 2030 in the absence of new policies.[18]

The continuing failure to address the energy needs of the poor threatens prospects for economic development. The transition from subsistence agricultural economies to modern industrial and service-oriented economies inherently requires adequate and affordable energy services. Moreover, reliance on traditional biomass directly threatens human health on a massive scale. Nearly half of all households around the world cook their daily food with unprocessed biomass (wood, coal, or dung). According to the World Health Organization, the result is deadly: "About 2.5 million deaths each year result from indoor exposures to particulate matter in rural and urban areas in developing countries, representing 4–5 percent of the 50–60 million global deaths that occur annually."[19]

For those reasons, the UN Millennium Project—an advisory body constituted by Kofi Annan, then the UN secretary general, to recommend practical steps toward achieving the Millennium Development Goals (MDGs)—argued forcefully that "greater quality and quantity of energy services will be required to meet the MDGs."[20]

Human Rights

The extractive industries in general suffer frequent accusations of gross human rights abuses. The UN's special representative on business and human rights, Harvard University professor John Ruggie, found that oil, gas, and mining firms "utterly dominate[d]" a survey of sixty-five instances of egregious human rights abuses, as reported by nongovernmental organizations (NGOs), accounting for two-thirds of the total.[21] The alleged abuses included such acts as "complicity in crimes against humanity, typically for acts committed by public and private security forces protecting company assets and property; large-scale corruption; violations of labor rights; and a broad array of abuses in relation to local communities, especially indigenous people."[22] As Ruggie notes, the predominance of extractive industries is no great surprise:

> No other [sector] has so enormous and intrusive a social and environmental footprint. At local levels in poor countries no effective public institutions may be in place. This authority vacuum may compel responsible companies, faced with some of the most difficult social challenges imaginable, to perform de facto governmental roles for which they are

all ill-equipped, while other firms take advantage of the asymmetry of power they enjoy.[23]

Many oil resources in particular are located in countries whose track record on human rights is less than stellar. The oil firms that do business in those countries find themselves tarred with the same brush. Shell Oil faced widespread criticism for alleged complicity in human rights violations in Nigeria in the 1990s.[24] Unocal found itself in U.S. federal court facing lawsuits over its alleged complicity in Myanmar, which it settled out of court in 2004.[25] Chinese oil companies in Darfur faced their turn in the spotlight after the turn of the century.[26]

In addition to claims of involvement in human rights abuses, advocacy groups have issued numerous reports documenting what they allege to be systematic complicity in misuse of government revenues from oil and gas by firms operating in repressive or poorly governed countries.[27] As one step toward countering such corruption, in 2002 the United Kingdom spearheaded the launch of the Extractive Industries Transparency Initiative (EITI), calling on governments that receive substantial energy revenues to publish accounts of those revenues.[28] The EITI puts the onus on governments, rather than corporations, to become transparent. Advocacy groups point out that very few governments have yet complied with the EITI requirement to publish fully audited and reconciled EITI reports.[29]

Governing Energy: Who Are the Governors?

Although there is no comprehensive global energy agency, a multitude of intergovernmental bodies and nongovernmental groups play some role in addressing global energy issues. They include large multilateral organizations that focus on energy, such as the IEA, the Energy Charter Treaty, and the International Energy Forum; a variety of small-scale, public-private partnerships and multi-stakeholder processes; bodies that focus on a specific energy source, such as the International Atomic Energy Agency; funders such as the multilateral development banks that include energy projects in their loan portfolios; and various business organizations, advocacy groups, and research institutions. A number of other institutions address energy as well, such as the G-8, the European Union (EU), and Asia Pacific Economic Cooperation (APEC). No single chapter can address more than a fraction of the entities involved. This one looks in depth at a few key organizations and governance processes to draw lessons about how global energy governance might, and should, evolve.

The International Energy Agency

Before 1974, no explicit agreements existed among governments to govern the actions of states and multinationals with regard to oil, the most easily transported and heavily traded energy source.[30] Then came the Arab-Israeli war of 1973. Members of the Organization of Petroleum Exporting Countries (OPEC) seized the opportunity to increase oil prices (and thus revenues) and simultaneously send a powerful political message by embargoing oil sales to countries that they considered overly friendly to Israel.[31] By December, global oil supplies had fallen by 7 percent.[32] Initially, oil-consuming nations responded competitively, in a manner uncomfortably reminiscent of the everyone-for-himself economic policies of the 1930s, when competitive devaluations and trade barriers turned a stock market crash into the most severe depression of the twentieth century. The nine-member European Economic Community (EEC) issued a pro-Arab resolution, which succeeded in easing Arab oil restrictions for those countries. However, Holland, which had maintained a publicly pro-Israeli stance, did not benefit from the easing of restrictions. Similarly, Portugal, while not a member of the EEC, suffered from oil restrictions because it had allowed the United States to use the Azores as a logistics base to channel U.S. weapons to Israel.[33] Many governments pressured oil companies to grant them priority in the allocation of available supplies. (The companies by and large declined to play favorites, instead allowing their customers to share the pain equally.)[34] The OECD secretariat proposed an oil-sharing arrangement to calm the panic, but to no avail.

In early 1974, the United States convened an international energy conference, at which the assembled governments agreed to create the International Energy Program, which established the IEA. By the end of 1978, the IEA was fully operational, housed at the OECD in Paris and comprising most OECD members (although not France). In what appeared to be a significant derogation of national sovereignty, the emergency oil-sharing system created under the agency's auspices delegated to the secretariat the authority to declare an emergency and thereby bring the system's operations into play. The agency also established systems for reporting on prices, supplies, and stock positions. Things seemed so tranquil in 1978 that many companies reduced their stockpiles of oil.[35]

The Iranian revolution destroyed that tranquility. With Iranian production down to almost nothing in early 1979, importing governments initially responded just as competitively as they had in the 1973 crisis. The scramble for supplies doubled prices, sparking a major increase in production that nonetheless failed to bring prices back down. The IEA secretariat never invoked the emergency oil-sharing system; instead, it attempted—unsuccessfully—to coordinate its members' actions informally.[36]

One IEA aim was to set oil import targets for each member. At a meeting in 1980, members approved a set of targets, but they were so high that they had no real effect on limiting demand. Over the next couple of years, IEA members tried but failed, despite strong U.S. support, to negotiate an agreement on a set of objective criteria by which the IEA could set national import targets. The debate over import targets proved useful as a way of bolstering the case for conservation efforts by keeping the need to control energy consumption on the agenda—but the failure to reach agreement showed the difficulty of getting governments to limit their sovereign autonomy for the greater good. IEA members were similarly unable to agree on a formal rules-based approach to managing and using oil stockpiles.[37]

The IEA also tried to set a minimum safeguard price (MSP) for crude oil. The MSP was supposed to be set at a level that would help stimulate investment in alternative energy sources—that is, to make investments in alternative energy sources financially worthwhile. The IEA's Standing Group on Long-Term Cooperation (SLT), which was in charge of energy conservation, development of alternative energy resources, and designing measures to reduce the dependence of member countries on oil, spent much of 1976 deliberating on the MSP.

However, the negotiation process was difficult, for multiple reasons. The interests of countries with alternative energy sources were different from those of countries without domestic reserves; the former countries wanted to maintain relatively high prices for imported oil in order to protect the public and private investments that went into developing their indigenous oil supply or alternative energy sources when imported oil prices declined.[38] There also were concerns about the effects of the MSP on industrial competition and advantage. Several countries argued that the United States and Canada had domestic legislation that allowed oil imports at prices below the MSP, which benefited their industries. Also at issue was chapter V of the Long-Term Cooperation Programme, which included the MSP and under which nondiscriminatory access by IEA members to others' indigenous energy resources was allowed. Several countries attempted to link their agreement to the MSP with concessions by oil-rich member countries, particularly concessions that allowed access to indigenous energy sources through joint energy projects, which Canada opposed. There also were disagreements on whether member countries would have to demonstrate that they had the administrative and legislative authority to maintain the MSP.[39] Nonetheless, an agreement finally was reached and the MSP was set to a relatively low price ($7 per barrel) in order to achieve consensus. In practice, however, the MSP was ignored and never gained the commitment of the IEA.[40] Furthermore, prices have since risen well above the $7 level, rendering the MSP grossly out of date.

Although the outbreak of the Iran-Iraq war in September 1980 drastically reduced available supplies, it had a much less dramatic effect on oil prices than did the Iranian revolution. Markets clearly played a major role in limiting the price impact—oil companies had turned their attention to the development of non-OPEC sources, and the depreciation of the dollar, in which oil prices are denominated, dampened price impacts.[41] The IEA also played a role in keeping oil markets calm. By then, the agency's reporting system was functioning and the secretariat was better placed to use its powers of persuasion on its members.[42] That helped to prevent the self-defeating cycle of stockpiling and hoarding that had characterized the earlier crises.

Since then, the IEA has helped to coordinate responses among consuming nations to a series of shocks and disruptions in global oil markets: the 1990–91 Gulf war; plans for dealing with Y2K concerns; 9/11; and the Iraq war. Throughout, the existence of the IEA and of its members' more than 1 billion barrels of Strategic Petroleum Reserve (SPR) helped to deter market manipulation.[43] In the United States, the Strategic Petroleum Reserve has been tapped twice. The first time was in 1991, the day before the Gulf war began, when an emergency sale of SPR crude oil was announced to ensure the adequacy of the global oil supply. The second time was in September 2005, after Hurricane Katrina devastated the oil production, distribution, and refining industries in Louisiana and Mississippi.[44]

The IEA also conducts energy research and compiles data, producing numerous publications on the latest energy statistics, policy analysis, and recommendations on good energy practices. These publications include the annual *World Energy Outlook* (*WEO*), which has become a leading source for energy market projections, analysis, and recommendations for governments and the energy business; regular reviews of the energy policies of individual member countries; oil market assessments; and monthly, quarterly, and annual energy statistics publications that serve as an important source of the information shaping the global energy debate. Also central to the work of the IEA is the fostering of innovation in energy technology, and the agency was asked at the 2005 G-8 Gleneagles Summit to make recommendations to that end. The IEA's subsequent work in the area resulted in the 2008 publication *Energy Technology Perspectives: Scenarios and Strategies to 2050,* which argued that technology was the key to sustainable energy development.[45]

However, serious questions have arisen about the system's capacity to cope with future shocks and disruptions. IEA membership is limited to countries that belong to the OECD. Its twenty-seven members now include all OECD countries except Iceland and Mexico. As oil demand soars among countries that are not members of either group, notably India and China, it is not clear that the agency has the critical mass of oil importers needed to manage a future shortfall.[46]

The Group of Eight

By any standard, the G-8 is an odd institution. With no charter, no permanent secretariat or home, no fixed membership, and no formal admission criteria, it nonetheless has become a fixture on the international scene, bringing together several of the world's most powerful leaders every year for more than thirty years. Although some analysts have come to denigrate the G-8 as nothing more than an inconsequential talking shop, over the years some of the G-8 summits appear to have helped to coordinate international action and establish norms. The G-8 has been especially active with regard to energy policy.[47]

The G-8, despite its name, began in 1975 with only six members (France, Germany, Italy, Japan, the United Kingdom, and the United States) at a summit in Rambouillet, France, initiated by France's president, Valéry Giscard d'Estaing. Canada joined the next year, creating the G-7. The European Community began participating in 1977. Russia took part in the political meetings in the early 1990s and became a full member in 1997.[48]

The G-8's attention to energy has waxed and waned, closely tracking oil prices.[49] In its early days (1975 to 1981), the G-7 did reasonably well in responding to the turmoil in oil markets. The Rambouillet declaration referred to the need to "cooperate in order to reduce our dependence on imported energy through conservation and the development of alternative sources" and proclaimed its signatories' commitment to "spare no effort in order to ensure more balanced conditions and a harmonious and steady development in the world energy market."[50] In subsequent years, member states made real and sometimes very detailed commitments. One paragraph of the 1978 Bonn Declaration, for example, was an extraordinarily public promise of specific U.S. policy measures:

> Recognizing its particular responsibility in the energy field, the United States will reduce its dependence on imported oil. The U.S. will have in place by the end of the year a comprehensive policy framework within which this effort can be urgently carried forward. By year-end, measures will be in effect that will result in oil import savings of approximately 2.5 million barrels per day by 1985. In order to achieve these goals, the U.S. will establish a strategic oil reserve of 1 billion barrels; it will increase coal production by two-thirds; it will maintain the ratio between growth in gross national product and growth in energy demand at or below 0.8; and its oil consumption will grow more slowly than energy consumption. The volume of oil imported in 1978 and 1979 should be less than that imported in 1977. In order to discourage excessive consumption of oil and to encourage the movement toward coal, the U.S. remains determined that the prices paid for oil in the U.S. shall be raised to the world level by the end of 1980.[51]

The communiqué of the 1980 Venice summit contained many pages of energy promises, couched in a nearly hysterical tone: "In this, our first meeting of the 1980s, the economic issues that have dominated our thoughts are the price and supply of energy. . . . Unless we can deal with the problems of energy, we cannot cope with other problems."[52] By 1982, however, oil prices were in decline. The G-8 was left in disarray by U.S.-European feuding over the proposed pipeline to bring natural gas from Russia's rich fields to energy-hungry Europe. Oil prices remained low through most of the next two decades, and despite continuing concern about nuclear proliferation, energy barely earned a mention in G-8 documents, other than a blip in the 1991 London communiqué due to the Gulf crisis.[53]

That began to change toward the end of the millennium. Japan, as host of both the 1997 Kyoto Protocol negotiations on climate change and the 2000 G-8 summit, wanted a strong new initiative on renewable energy.[54] At its Okinawa summit in 2000, the G-8 tried something new, creating the G-8 Renewable Energy Task Force, which was co-chaired by Sir Mark Moody-Stuart, the head of Shell Corporation, and Corrado Clini, director general of Italy's Department of Environment. Its membership drew not only from G-8 governments but also from business and civil society and from non–G-8 countries.[55] The task force report, delivered to the G-8 in July 2001, laid out a compelling case for a major shift to sources of renewable energy and set out recommendations for using market forces and a variety of funding mechanisms to bring about that shift.[56]

The work of the task force has to go down as one of the major missed opportunities for getting the world onto a more sensible and sustainable energy path. By 2001 the political landscape of the G-8 had changed dramatically. George W. Bush, at his first G-8 summit, seemed to see the task force's work as a Clintonian exercise of no interest to the incoming administration. The 2001 Genoa summit barely acknowledged the report and let the task force die.[57]

By 2004 rising oil prices and the perceived connection between Middle East oil revenues and economic vulnerability and terrorism helped turn G-8 leaders' attention back toward energy policy. By 2005, with climate change at the top of host Tony Blair's agenda, energy policy featured in much of the discussion at the Gleneagles summit, with serious commitments on energy efficiency, cleaner energy technology, and investment in such technologies for developing countries. Given the publicity generated over the G-8 energy commitments, there was much activity by all member states on climate change and renewable energy initiatives.[58] All member states had participated at the UN Climate Change Conference in November 2005 and had accepted more than forty key agreements. The most significant was the adoption of the 2001 Marrakesh Accords, which established how many of the Kyoto Protocol's mechanisms would be adopted, and an agreement to move forward on post-2012 emissions reduction negotiations.[59] The member states also adopted various domestic legislative plans to foster renewable energy initiatives. For instance, France announced Plan Climat 2005, which committed France to develop markets in clean energy technology and to increase the availability of the technology in developing countries. France also promised tax credits to private individuals who repurchased electricity from solar panels.[60] The United Kingdom set a target goal to achieve 15.4 percent of its energy from renewable sources by 2015–16. And the United States announced the Advanced Energy Initiative, which increased funding for clean energy research at the Department of Energy by 22 percent.[61]

The St. Petersburg 2006 G-8 summit hosted by Russia had energy as its central focus. Ironically, the Russian state-owned gas supplier Gazprom had begun 2006 by shutting off the gas pipes to Ukraine over a price dispute. In doing so, Russia violated its obligations under the Energy Charter Treaty, of which it was a signatory but which it had not ratified (certain obligations being incumbent on signatories even as they consider ratification).

A few months before the St. Petersburg summit, John Kirton, a leading Canadian authority on the G-8 (and a strong proponent of the view that the G-8 has been quite successful on energy policy), presented a Moscow conference with a set of eminently sensible recommendations on what the G-8 should do that summer to take advantage of its combined political and economic muscle and its past successes in the energy field. He suggested that the summit focus intensively on energy, framed as "environmentally sustainable energy," with particular emphasis on mobilizing the market to carry out whatever specific commitments the G-8 would make. Those commitments would include serious attention to rebalancing subsidies, away from nuclear and the dirtier fossil fuels and toward cleaner and more sustainable sources; a shift toward ecological national accounting that would reveal the real costs of existing energy policies; creation of a more global natural gas market using LNG; and greater institutionalization of G-8 energy institutions, at the ministerial, official, and multistakeholder levels.[62]

The G-8 leaders did none of those things. The summit's Global Energy Security statement was long on what should be done but vague on what exactly the G-8 governments would themselves do. It was silent on subsidies and ecological accounting. Although the statement treated energy efficiency, renewable energy, and new energy technology at length, the most concrete commitment was to "consider national goals for reducing energy intensity of economic development, to be reported by the end of the year."

At the 2007 G-8 Summit at Heiligendamm, climate change received passing attention with an acknowledgment of a report by the UN Intergovernmental Panel on Climate Change (IPCC) that had concluded that global temperatures were rising because of human activity and that the rise would result in negative consequences for biodiversity and ecosystems (food and water supply). The G-8 also acknowledged an EU proposal for an international initiative on energy efficiency and agreed to explore, along with the International Energy Agency, ways to promote global energy efficiency. Still, no firm plan for action was presented, only a declaration to "consider" halving global emissions by 2050 and a call for members to participate in the UN process for negotiating a post-Kyoto agreement.[63]

At the 2008 G-8 Summit at Hokkaido Toyako, the leaders made a commitment to adopt in negotiations on the UN Framework Convention on Climate Change (UNFCCC) the goal of achieving at least a 50 percent reduction of global emissions by 2050. Topics such as improving energy efficiency, use of clean energy, adaptation to the effects of climate change, technology, finance, market-based mechanisms, and tariff reduction were discussed. The G-8 countries, along with China, India, South Korea, and the European Community, established the International Partnership for Energy Efficiency Cooperation. The group made a commitment to increase investment in research and development in innovative energy technology, and to achieve that end members pledged more than US$10 billion annually in direct government-funded R&D over the next several years. G-8 members also pledged approximately US$6 billion as a contribution to the Climate Investment Funds, including the Clean Technology Fund (CTF) and the Strategic Climate Fund (SCF).[64]

Although the G-8 process may have helped to focus the attention of great powers on energy issues, to date the process has not proven its ability to serve as a central mechanism for global energy governance.

The Energy Charter Treaty

The end of the cold war seemed to offer a new opportunity to bolster energy markets and thus energy security by incorporating at least one major supplier—Russia—into a rules-based framework. In December 1991, following Dutch prime minister Ruud Lubbers's proposal for a European Energy Community, a number of European countries signed the Energy Charter, a political declaration of intent to promote cooperation on energy issues. That led three years later to the Energy Charter Treaty (ECT), signed in Lisbon in December 1994, with entry into force in April 1998 upon ratification by thirty members.[65] Membership now stands at fifty-one countries plus the European Communities, including a number of non-European parties such as Australia, Japan, and all Central Asian countries. Countries and organizations with observer status include China, the United States, Venezuela, Iran, Kuwait, ASEAN, the World Bank, the OECD, the IEA, and the Commonwealth of Independent States (CIS) Electric Power Council, among others. The United States signed the ECT but chose not to ratify it.

The ECT, which started off being inherently European, has become more or less global in its scope, in line with its current ambitions.

Although the ECT includes attention to energy efficiency as one of its five pillars, the ECT's purpose, as its website makes clear, is to stabilize markets and thus enhance energy security:

> In a world of increasing interdependence between net exporters of energy and net importers, it is widely recognized that multilateral rules can provide a more balanced and efficient framework for international cooperation than is offered by bilateral agreements alone or by non-legislative instruments. The Energy Charter Treaty therefore plays an important role as part of an international effort to build a legal foundation for energy security, based on the principles of open, competitive markets and sustainable development.[66]

The ECT's other four pillars address foreign energy investment, energy trade, freedom of transit through pipelines and grids, and dispute resolution. On investment, under the terms of the treaty, each party is obliged to extend national treatment (most-favored-nation status) to nationals and legal entities of all other parties that have invested in its energy sector, thus replacing the need for a network of bilateral investment protection treaties. On trade, the ECT accepts World Trade Organization (WTO) rules and standards, extending WTO-type rules to several ECT parties that are not yet members of the WTO.[67] The dispute resolution procedure relies on arbitration.

It is above all the transit issue that has proven problematic, particularly for Russia. In 1998, a number of Russia's energy-exporting neighbors and transit countries (Azerbaijan, Georgia, Kazakhstan, Turkey, Turkmenistan, and Uzbekistan) raised the issue, arguing that if commercial oil and natural gas pipeline projects were to succeed, an attractive political, technical, financial, and legal environment would have to be created. Since the pipelines cross borders, creating an attractive commercial environment would require an intergovernmental agreement. Attendees at the G-8 Energy Ministerial Meeting that year agreed and established the Transit Working Group. Negotiations on a transit protocol began under ECT auspices in early 2000.[68]

Russia, however, proved unwilling to agree to provisions that in effect allowed non-Russian companies to buy gas in Central Asia and ship it to Europe through Russian pipelines, instead of having to sell it to Russia, which would then convey the gas to Europe. In late 2006, Russia made clear that it did not intend to give up control of its pipelines and would not ratify the Energy Charter Treaty unless those provisions were renegotiated.[69] It also seems likely that Russia does not wish to submit to the ECT's arbitration procedures for price disputes and its ban on cutting off supplies.[70]

The difficulties over the ECT are just one piece of a larger global governance failure. The effort to incorporate Russia into a rules-based energy market system has failed spectacularly. Flush with cash and confidence, Russia has bullied foreign energy firms out of the enormous Sakhalin Island project and has cut off supplies to Ukraine and Belarus in pricing disputes.[71] In October 2006, Gazprom reversed a major policy decision, announcing that it would develop the enormous Shtokman gas fields without the foreign investors who previously were to have been allocated a 49 percent share.[72]

The WTO

As many energy exporters, including OPEC member countries, Central Asian countries, and Russia, negotiate the terms of accession for their entry into the World Trade Organization, the WTO is taking on increasing importance as a focal point for energy-relevant trade rules. Trade rules cover most of the policy instruments that governments have to improve energy efficiency and govern their energy sectors, from taxation to subsidies to standards and labeling requirements.[73]

Trade rules fit awkwardly with energy policy. WTO rules are meant to address import barriers—tariffs and other measures that countries use to keep out other countries' goods and services. With regard to energy, however, few import barriers exist. Most energy importers are scrambling to increase imports, not exclude them. Instead, the barriers to trade come from exporters, in such forms as export duties, which can raise significant revenues for exporting countries. WTO rules do not address supply monopolies or cartels or issues such as the pipeline transit rules that have derailed Russia's ratification of the Energy Charter Treaty.

In many cases, WTO rules may inhibit good energy policy. Carbon taxes on fuels, which several countries already have adopted, would pass muster.[74] It is not clear, however, whether the rules would allow tax policy to discriminate between methods of energy production, such as favoring electricity from renewable sources over electricity from other sources. Similarly, direct support to renewable energy industries may fall afoul of WTO prohibitions on subsidizing specific industries within a sector.[75] In the meantime, perverse but long-established subsidies—which benefit greenhouse gas–producing energy sources at the expense of cleaner ones— abound. The world's poorer countries (non-OECD members) subsidize oil products to the tune of more than $90 billion a year.[76]

It has been argued that environmental principles and perspectives have progressed in the WTO through cases brought to the Dispute Settlement Body (DSB) and the Appellate Body. The appellate process is supposed to help to clarify and interpret WTO law and fill the gaps left by negotiators.[77] However, it is not always clear that trade or investment agreements will help foster clean energy initiatives. In fact, bilateral and regional investment agreements, which have considerably stronger provisions than the WTO agreements do, can actually restrict the effectiveness of government policies that favor clean energy initiatives over less environmentally friendly ones. Such agreements can be open to two types of restrictions, one regarding expropriation and the other regarding fair and equitable treatment. With the former, if a government's new clean energy policy had a significantly negative financial impact on an investment made by a foreign investor, then the investor could argue that his investment was being indirectly expropriated and claim damages. The second restriction simply amounts to unchanged regulatory treatment for the investor; if new environmental regulations adversely affect the finances of a business, that might be a basis for arbitration under the obligation to provide fair and equitable treat- ment.[78] Therefore, because trade or investment agreements are open to such challenges, they are insufficient to address the issue of fostering good energy governance.

International Energy Forum

Since its first meeting in 1991 in South Africa, the International Energy Forum (IEF) has brought together the energy ministers of energy-producing and -importing countries every year to exchange views.[79] Based in Saudi Arabia, it focuses almost exclusively on oil and natural gas, and does not address energy security, diversification, renewable energy, or environmen- tal issues. Its major accomplishment to date is the establishment of the Joint Oil Data Initiative (JODI) to improve the availability and reliability of international data on crude oil, LPG, gasoline, kerosene, gas/diesel, heavy oil, and so forth.[80] As of November 2007, ninety-seven countries were participating, but JODI was still a work in progress.

The World Bank

Global understanding of the strong connection between energy and development is growing. Over the course of the various global environment and development summits, from the 1972 Stockholm Summit to the 2002 Johannesburg World Summit on Sustainable Development, the topic of energy became ever more prominent.[81] Yet providing adequate financing and an appropriate policy framework for energy in developing countries remains problematic, and an especially egregious shortfall exists in funding to meet the needs of the poorest people. Although most financing for energy development comes from the private sector, various agencies of the UN system and the multilateral development banks, in particular the World Bank, play a key role in setting the terms of the debate and in providing funding.

At the Gleneagles Summit in 2005, the G-8 asked the World Bank to take a leadership role in creating a new framework for clean energy and development, including investment and financing.[82] At the 2006 annual meeting of the IMF/World Bank Board of Governors, held in Singapore that year, the bank released its strategy.[83] The report acknowledged that meeting the MDGs would require far more aggressive action than is contemplated under the IEA's "business as usual" scenario. To address the needs of the poorest, the report called for an action plan that included five components and gave particular attention to sub-Saharan Africa:

—scaled-up programs of household electrification (with better integration of mini-grid and off-grid options to complement grid-based approaches)

—additional generation capacity with associated transmission capacity (including through regional projects) to serve newly connected households and demand from enterprises, public facilities, and other users

—access to clean cooking, heating, and lighting fuels (through sustainable forest management, fuel switching, and diffusion of improved charcoal, briquetting, and clean cooking technologies)

—provision of energy services for key public facilities, such as schools and clinics

—provision of stand-alone lighting packages for households without electricity service.[84]

However, the financing that would be needed to connect all households for electricity by 2030 is substantial, on the order of $34 billion a year through 2030.[85]

With regard to making the transition to clean energy, the World Bank's report pointed out that despite the existence of some funding vehicles at the various multilateral development banks (the European Bank for Reconstruction and Development's Energy Efficiency Facility, the Inter-American Development Bank's Sustainable Energy Initiative, and the Asian Development Bank's Asia Pacific Energy Efficiency Fund, in addition to the World Bank's funding), the funding available for clean energy projects was negligible compared to the need. It called for the establishment of a new Clean Energy Financing Vehicle (loans) and a new Clean Energy Support Vehicle (grants).

The report was not well received. Developing country governments were highly suspicious of the initiative because of its source—the G-8 and the North-dominated World Bank. The environmental and development research and activist communities claimed that the strategy failed to serve the interests of the poor or to make serious progress toward limiting greenhouse gas emissions. They argued that the bank's proposed investment framework reflected a longstanding predisposition at the bank to use its funding to support a business as usual. In a report also issued at the Singapore meetings, bank critics made a series of pointed criticisms of the bank's approach to energy.[86] Overall, as the critics point out, the bank's strategy—and indeed its lending portfolio—does little more than tinker around the edges, largely accepting the continuation of business as usual in energy rather than trying to lead the way to a more fundamental transformation of the energy sector. Although the bank, under pressure from NGOs, has repeatedly promised significant changes in its policies,[87] its actual practices remain firmly wedded to lending for centralized large-scale and mostly fossil fuel–based energy projects, to the tune of $2 to $3 billion a year, some ten times the amount made available for other energy sources. As figure 7-2 indicates, both the World Bank and the International Finance Corporation devote very little of their loan portfolios to renewable sources of energy. The bank, for its part, contends that renewables are not available on the scale needed to meet the world's growing demand and that continued investment in fossil fuel projects in poor countries is essential.

Other Initiatives

At the 2002 Johannesburg World Summit on Sustainable Development and thereafter, a wide range of initiatives have attempted to redress some of the shortcomings of energy governance. Many of them focus on renewable

FIGURE 7-2. World Bank and IFC Energy Funding

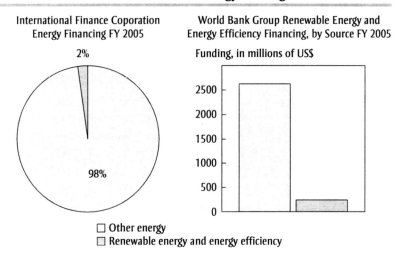

International Finance Coporation Energy Financing FY 2005

World Bank Group Renewable Energy and Energy Efficiency Financing, by Source FY 2005

☐ Other energy
☐ Renewable energy and energy efficiency

Source: *Friends of the Earth, Power Failure: How the World Bank Is Failing to Adequately Finance Renewable Energy for Development*, p. 13 (www.foe.org/camps/intl/institutions/renewableenergyreport10242005.pdf).

sources of energy, reflecting the view among many environmentalists and some development specialists that renewables provide a double whammy— avoiding greenhouse gas emissions and other environmental externalities and often providing local jobs and more easily decentralized energy sources.[88] A few of the more notable undertakings include the following:

—The EU Energy Initiative for Poverty Eradication and Sustainable Development, launched at Johannesburg, helps developing countries maximize energy efficiency and increase the use of renewable sources of energy.[89]

—The London-based Global Village Energy Partnership, also launched in Johannesburg, aims to help developing countries establish energy action plans and brings together some 1,500 small and medium enterprises (SMEs) and NGOs involved in energy in developing countries with donors and providers of technical assistance.[90]

—The Renewable Energy and Energy Efficiency Partnership (REEEP), launched by the United Kingdom in August 2003 as a multi-stakeholder coalition to promote use of renewable resources and energy-efficient systems, works on policy and regulatory initiatives for clean energy and facilitates financing for energy projects, with the backing of more than 200 national governments, businesses, development banks, and NGOs. It has eight regional secretariats around the world, in addition to the international secretariat.[91]

—REN21 (Renewable Energy Policy Network for the 21st Century) grew out of the Renewables 2004 conference in Bonn. Its thirty-two-member steering committee includes representatives from governments, intergovernmental organizations (IGOs), NGOs, industry, finance, regional governments, local governments, and members at large. With a Paris-based secretariat, REN21 hosts meetings, issues publications, and broadly advocates for good renewable energy policies.[92]

—The Asia-Pacific Partnership for Clean Development and Climate was established by Australia, China, India, Japan, the Republic of Korea, and the United States to accelerate the development and deployment of clean energy technologies and related goods and services. The partnership has set up eight public-private task forces covering aluminum, buildings and appliances, cement, cleaner use of fossil fuel, coal mining, power generation and transmission, renewable energy and distributed generation, and steel. The members represent roughly half of the world's economy, population, and energy use.[93]

Building Blocks for Energy Governance

The challenge for global energy governance is daunting, given the massive scale of the problem. The IEA's forecast of the need for $45 trillion in new energy investment by 2050 is almost certainly a gross underestimate, because the IEA does not assume that the world will provide full access to energy services to the world's poor, does not fully account for the costs of adequately protecting the environment (reducing CO_2 emissions by 50 percent from the current levels is only a beginning) or human rights, and does not take into account the costs of protecting energy infrastructure. As the preceding analysis makes clear, existing governance mechanisms are failing to provide energy security, address energy-related environmental externalities, protect human rights from violations during the process of extracting energy resources, or ensure that energy services are sufficiently available to the poor to meet the MDGs and other development goals. However, it is not easy to put forward feasible recommendations for making significant improvements in the processes of global energy governance.

New or Expanded Intergovernmental Organizations?

A common response to perceived needs for global governance is to call for the creation or expansion of a formal intergovernmental organization, preferably one with teeth. For example, in February 2007 Jacques Chirac, then the president of France, called for the transformation of the UN's Environment Programme into a "genuine international organization to which all countries belong, along the lines of the World Health Organization," to promote sustainable technologies and behavior patterns and to support "the implementation of environmental decisions across the planet"[94]—a proposal that obviously would have significant implications for energy policy. Similarly, it frequently has been proposed that the International Energy Agency, as the club of major oil importers, should expand its membership to include at a minimum China, India, and other emerging markets. The purpose of creating the International Energy Forum, which unlike the IEA includes most oil exporters as well as importers, was to pull together all parties on energy issues.

Nonetheless, the near-term prospects for new overarching formal organizations or for substantial expansion of their authority are not bright. As one recent analysis of global governance concluded:

> [T]he conditions at the beginning of the twenty-first century do not seem ripe for any major systemic breakthroughs that would replace current structures and create new institutions. The vision and sense of urgency, the innovative spirit, and the leadership that brought the IMF and the World Bank into being at Bretton Woods in 1944 and created the United Nations in San Francisco in 1945 are not present today.[95]

It is not surprising, therefore, that Chirac's repeated calls for a "World Environmental Organization" have not been strongly endorsed by other major powers. Although more than forty countries supported the proposal, the United States, India, China, and Russia all expressed opposition.

Even the prospect of expanding the IEA seems uncertain. Aside from the membership criterion of democratic governance (an artifact of the IEA's origin as an OECD creation), there are serious worries over sharing data, doubts about the capacity of China and India to meet the basic requirement to create and maintain a ninety-day oil stockpile, and concerns about disruption of the IEA's internal political balance. The IEA has what must be the most convoluted voting structure of any intergovernmental organization—one published effort to explain it runs sixteen pages—but the important point is that the system is carefully balanced to ensure that decisions on most issues require either unanimity or special majorities and that neither the United States nor the EU is in a position to veto a decision requiring a majority vote.[96] Because voting weights are calculated in part on the basis of oil consumption, the addition of India and China would make those countries' voting shares equal to or ahead of those of all other members except the United States.

Intermediate Steps toward Global Energy Governance

In an ideal world, energy governance, like all forms of global governance, would entail fully accountable institutions with widespread participation, able to supply the full range of energy-related public goods. A starting point would be to recognize the connected nature of all the items on the full energy agenda. To that end, it might be useful to take each one of these difficult and intractable big problems and make them bigger. Creating a more coherent framework would have the great advantage of allowing for grand bargains that could ensure that everyone's most fundamental interests were met.

It may, for example, be more effective to bundle climate change with broader energy issues than to treat it in isolation. It is much easier to make a case for why both rich and poor countries should adopt sound energy policies for reasons of geostrategic, environmental (including non–climate change environmental), and developmental self-interest than it is to persuade developing nations that they should bear a significant part of the burden of countering a climate change problem that they had little part in creating. There is at least some hope that such major players as the United States and China might be receptive to such a broader approach. During his 2008 election campaign, U.S. president Barack Obama outlined his New Energy for America plan. In addition to other targets, his plan would implement an economy-wide cap-and-trade program to reduce greenhouse gas emissions 80 percent by 2050 and would invest $150 billion over ten years to foster private sector efforts to build a clean energy future.[97] That would make it more difficult for China to use U.S. inaction as an excuse for its own nonparticipation in global governance efforts. China's public statements on climate change indicate a rapidly growing awareness among Chinese leaders of the potentially disastrous impacts on China from global warming.

Energy policy already is a central focus in such overarching institutions as the G-8 and the EU. With real political leadership, from the United States in particular, a broad global consensus on a more coherent approach to energy is not out of the realm of possibility, but that is a goal that will require years of dedicated effort to achieve. And if it is ever to be achieved, it can be implemented only if the capacity for international collaboration on energy is greatly strengthened. It is more likely that, at least in the short term, improvements in energy governance will be piecemeal and incremental. Nonetheless, such improvements could make a real, if limited, difference. Even those steps, however, will require patience and leadership.

—The World Bank could put its funding where its rhetoric has long been by giving much more attention to energy efficiency and to the possibility of a massive, bank-funded shift to renewable energy technologies: wind, solar, modern biomass, geothermal, and small hydropower. To do so would require strong consensus among the bank's member countries and strong support from the World Bank's president.

—Existing IEA outreach efforts to China and India could be expanded to include the development of a more global system of reserves and emergency stocks.

—The Joint Oil Data Initiative could be developed further, serving as a prototype for other systems to provide timely and accurate information on global energy markets. Further development would require a concerted effort to overcome the reluctance of many oil and gas producers to release accurate data on reserves, which many of them treat as national security information.

—More effective diplomacy could help to entice Russia into a more constructive role in the Energy Charter Treaty and other international energy governance regimes. The rapid decline in oil prices in the second half of 2008 opens a (probably temporary) window of opportunity to engage Russia on this subject.

—Expanded political support from both governments and corporations for the Extractive Industries Transparency Initiative could reduce what currently are extremely high levels of corruption associated with the extraction of energy resources.

These proposals still leave enormous gaps. There remains a need for full, timely, and accurate information on the environmental externalities resulting from various energy policies. There remains a need for a globally agreed system to develop redundant and therefore resilient infrastructure (pipelines, refineries, decentralized supply, and so forth) for providing energy services, a need that will almost certainly go unmet given the current economic downturn. Most important, there is a pressing need for a meeting of minds to ensure that energy is conceived of as a shared interest rather than an object of geopolitical competition.

If we accept the premise that energy is a zero-sum game, there is little room for optimism that global governance can cope. However, if we redefine the energy problem to focus not on particular sources of energy (such as oil) but rather on energy services, the picture is somewhat rosier. Most of the world has a mutual interest in developing effective energy markets, coordinating policies on taxes and subsidies, responding effectively to climate change, and making a serious investment in alternative energy technologies for developing countries in order to put them on a sustainable path now rather than later through retrofitting. These are all obvious policy prescriptions, repeated in numerous reports. What has been lacking to date is a concerted effort to develop better habits of collaboration, to learn from experience, and to develop the organizational infrastructure needed for progress. The world will not be able to shift to energy systems that satisfy the demands for reliable access to energy services for all the world's people at an environmentally acceptable cost unless the international community develops far more effective institutional mechanisms.

Notes

1. Intergovernmental Panel on Climate Change Fourth Assessment Report, released February 2, 2007 (www.ipcc.ch/).

2. The IEA report presented several different scenarios, from a technological perspective, on the feasibility and costs of deep emission reductions. International Energy Agency, *Energy Technology Perspectives 2008: Executive Summary* (Paris: OECD/IEA, 2008), pp. 2–3.

3. This list is drawn from Frances Seymour and Simon Zadek, "Governing Energy: The Global Energy Challenge," *AccountAbility Forum* 9 (2006), pp. 6–15.

4. Daniel Yergin, "Ensuring Energy Security," *Foreign Affairs*, March–April 2006.

5. Charles Maechling, "Pearl Harbor: The First Energy War," *History Today* (December 2006).

6. Cited in Susan Strange, *States and Markets*, 2nd ed. (London: Continuum International Publishing Group, 1998), p. 201.

7. Manjeet Singh Pardesi and others, "Energy and Security: The Geopolitics of Energy in the Asia-Pacific" (Singapore: Institute of Defense and Strategic Studies, October 2006), p. 5.

8. Yergin, "Ensuring Energy Security."

9. For a good discussion of energy markets, see the interview with several Stanford faculty members in "A Crude Awakening," *Stanford Magazine* (November–December 2006) (www.stanfordalumni.org/news/magazine/2006/novdec/features/energy.html).

10. David Victor, cited in "A Crude Awakening."

11. Jonathan H. Adler, "Let Fifty Flowers Bloom: Transforming the States into Laboratories of Environmental Policy," paper prepared for American Enterprise Institute, Federalism Project Roundtable on Federalism and Environmentalism, September 20, 2001, p. 44.

12. Benjamin K. Sovacool, "The Best of Both Worlds: Environmental Federalism and the Need for Federal Action on Renewable Energy and Climate Change," *Stanford Environmental Law Journal* 27, no. 2 (June 2008), p. 472.

13. Intergovernmental Panel on Climate Change Fourth Assessment Report.

14. See the chart at http://seawifs.gsfc.nasa.gov/OCEAN_PLANET/IMAGES/spills_chart.gif.

15. Food and Water Watch and Network for New Energy Choices, *The Rush to Ethanol: Not All Biofuels Are Created Equal. Analysis and Recommendations for U.S. Biofuels Policy* (New York: 2007), p. 7.

16. For an excellent analysis, see *Dams and Development: A New Framework for Decisionmaking,* World Commission on Dams, released November 16, 2000.

17. Benjamin K. Sovacool, *The Dirty Energy Dilemma: What's Blocking Clean Power in the United States* (Westport, CT: Praeger Publishers, 2008), p. 119.

18. IEA, *World Energy Outlook 2008,* p. 178 (www.worldenergyoutlook.org/2008.asp).

19. Nigel Bruce, Rogelio Perez-Padilla, and Rachel Albalak, "The Health Effects of Indoor Air Pollution Exposure in Developing Countries," WHO/SDE/OEH/02.05 (World Health Organization, 2002), p. 5.

20. Vijay Modi and others, "Energy Services for the Millennium Development Goals" (Washington and New York: The World Bank and the United Nations Development Programme, 2005), p. 2.

21. John Ruggie, "Interim Report of the Special Representative of the Secretary-General on the Issue of Human Rights and Transnational Corporations and Other Business Enterprises," U.N.DocE/CN.4/2006/97 (United Nations, 2006), p. 5.

22. Ibid.

23. Ibid., p. 6.

24. Clifford Bob, *The Marketing of Rebellion: Insurgents, Media, and International Activism* (Cambridge University Press, 2006), pp. 54–56, available through Yale Global Online.

25. Ashild Kolas and Stein Tonnesson, *Burma and Its Neighbors: The Geopolitics of Gas,* Austral Policy Forum 06-30A, August 24, 2006 (http://nautilus.rmit.edu.au/forum-reports/0630a-kolas-tonnesson.html).

26. Human Rights Watch (www.hrw.org/reports/2003/sudan1103/8.htm#_Toc54492553); Amnesty International USA (www.amnestyusa.org/countries/china/document.do?id=ENGAFR540722006).

27. See, for example, the various reports by Human Rights Watch at http://hrw.org/doc/?t=corporations-extract; Global Witness, *Time for Transparency: Coming Clear on Oil, Mining, and Gas Revenues,* March 2004 (www.globalwitness.org/media_library_detail.php/115/en/time_for_transparency).

28. See EITI website (www.eitransparency.org/).

29. *Eye on EITI: Civil Society Perspectives and Recommendations on the Extractive Industries Transparency Initiative* (London and New York: Publish What You Pay and Revenue Watch Institute, 2006).

30. The best academic analysis of the early days of the IEA, from which the following paragraphs draw, is Robert Keohane, *After Hegemony: Cooperation and Discord in the World Political Economy* (Princeton University Press, 1984), pp. 217–40.

31. Strange, *States and Markets*, p.197.

32. Keohane, *After Hegemony*, p. 222.

33. Ibrahim Sus, "Western Europe and the October War," *Journal of Palestine Studies* 3, no. 2 (Winter 1974), pp. 70–73.

34. Keohane, *After Hegemony*, p. 223.

35. Ibid., pp. 224–26.

36. Ibid., p. 230.

37. Ibid., pp. 233–34.

38. Mason Willrich and Melvin A. Conant, "The International Energy Agency: An Interpretation and Assessment," *American Journal of International Law* 71, no. 2 (April 1977), p. 212.

39. Robert O. Keohane, "The International Energy Agency: State Influence and Transgovernmental Politics," *International Organization* 32, no. 4 (Autumn 1978), p. 940.

40. Ian Skeet, OPEC: *Twenty-Five Years of Prices and Politics* (CUP Archive, 1991), p. 125.

41. Strange, *States and Markets*, p. 198.

42. Keohane, *After Hegemony*, pp. 235–36.

43. *Strategic Energy Policy Challenges for the 21st Century*, Report of an Independent Task Force Cosponsored by the James A. Baker III Institute for Public Policy of Rice University and the Council on Foreign Relations (2001) (www.informationclearing house.info/article3535.htm).

44. For more information, see U.S. Department of Energy, "Releasing Crude Oil from the Strategic Petroleum Reserve" (http://fossil.energy.gov/programs/reserves/spr/spr-drawdown.html).

45. For further details on the IEA's work, see "About the IEA" (www.iea.org/about/docs/iea2008.pdf).

46. Strategic Energy Policy, p. 33.

47. See, for example, Colin Bradford and Johannes Linn, "The Irrelevant G-8 Summit in St. Petersburg," Brookings op-ed (www.brookings.edu/views/op-ed/200607 irg8.htm).

48. For information on the history of the G-8, see the various official websites of the host countries, for example, "Official Website of the G-8 Presidency of the Russian Federation in 2006" (http://en.g8russia.ru/g8/history/shortinfo/index-print.html). For comprehensive information on the G-8, see the website of the G-8 Information Centre at the University of Toronto (www.g7.utoronto.ca/).

49. For a good overview of the G-8's energy performance through the 2005 Gleneagles summit, see John Kirton, "The G-8 and Global Energy Governance: Past Performance, St. Petersburg Opportunities," paper presented at "The World Dimension of Russia's Energy Security," conference sponsored by the Moscow State Institute of International Relations (MGIMO), Moscow, April 21, 2006 (www.g8. utoronto.ca/scholar/kirton2006/kirton_energy_060623.pdf). For a helpful compi-

lation of all G-7/G-8 documents on energy, see John Kirton and Laura Sunderland, "The G-8 Summit Communiqués on Energy, 1975–2005" (www.g8.utoronto.ca/references/energy.pdf)

50. 1975 Rambouillet Declaration, paragraph 13, in Kirton and Sunderland, "The G-8 Summit Communiqués on Energy," p. 2.

51. See "Energy," G-7 Summit, Bonn, July 16–17, 1978 (www.g7.utoronto.ca/summit/1978bonn/communique/energy.html).

52. Kirton and Sunderland, "The G-8 Summit Communiqués on Energy," p. 9.

53. Ibid., pp. 20–23.

54. Kirton, "The G-8 and Global Energy Governance," p. 11.

55. See the Renewable Energy Task Force (www.g8.utoronto.ca/meetings-official/g8renewables_report.pdf).

56. G-8 Renewable Energy Task Force, "Chairman's Report," July 21, 2001 (www.g8.utoronto.ca/meetings-official/g8renewables_report.pdf).

57. Kirton, "The G-8 and Global Energy Governance," p. 11.

58. John Kirton and others, "2005 Gleneagles Final Compliance Report," June 12, 2006 (www.g7.utoronto.ca/evaluations/2005compliance_final/index.html). The report reviews the progress made on selected commitments set out at the 2005 G-8 Gleneagles Summit.

59. Ibid., p. 203.

60. Often private homes with solar panels are also connected to the commercial electric grid because the panels do not always produce energy—for example, during the night or when it rains. The commercial grid supplements the panels when they cannot supply the home's electricity needs. When the solar panels produce more electricity than is needed, the excess is sold to the electric company. Repurchasing occurs when a household buys back the amount of electricity that it sold to the electricity company at a later date, when it needs the energy most.

61. Ibid., pp. 187–98.

62. Kirton, "The G-8 and Global Energy Governance," pp. 22–25.

63. Documents pertaining to the outcome of the 2007 G-8 Heiligendamm Summit are available at www.g-8.de/Webs/G8/EN/G8Summit/SummitDocuments/summit-documents.html.

64. Documents pertaining to outcome of the 2008 G-8 Summit at Hokkaido Toyako are available at www.mofa.go.jp/policy/economy/summit/2008/doc/doc080714__en.html.

65. See "About the Charter" (www.encharter.org/index.php?id=7).

66. Ibid.

67. See "Trade and Transit" (www.encharter.org/index.php?id=5&L=0).

68. See "Transit Protocol" (www.encharter.org/index.php?id=37).

69. Judy Dempsey, "Russia Gets Tough on Energy Sales to Europe," *International Herald Tribune*, December 12, 2006 (www.iht.com/articles/2006/12/12/news/energy.php [February 2007]).

70. Novosti Russian News and Information Agency, "Russia Would Ratify Amended Energy Charter Treaty—Kremlin Aide," April 26, 2006 (http://en.rian.ru/russia/2006 0426/46936292-print.html).

71. "After Sakhalin; Russian Energy," *The Economist*, December 16, 2006, pp. 75, 13.

72. Dempsey, "Russia Gets Tough on Energy Sales to Europe."

73. Malena Sell, "Trade, Climate Change, and the Transition to a Sustainable Energy Future: Framing the Debate," in *Linking Trade, Climate Change, and Energy*, ICSTD Issue Brief (International Centre for Trade and Sustainable Development, November 2006), p. 1 (www.icstd.org).

74. Finland, Italy, Norway, the Netherlands, Sweden have national carbon taxes, and New Zealand is expected to adopt such a tax. International Task Force on Global Public Goods, *Meeting Global Challenges: International Cooperation in the National Interest* (2006), p. 114 (www.gpgtaskforce.org).

75. Yulia Selivanova, "Transition to a Sustainable Energy Future: Global Trade Rules and Energy Policies," in *Linking Trade, Climate Change, and Energy*, pp. 3–4.

76. IEA *World Energy Outlook 2006*, p. 39 (www.worldenergyoutlook.org/2006.asp).

77. Mark Halle, "The WTO and Sustainable Development," in *The WTO in the Twenty-First Century: Dispute Settlement, Negotiations, and Regionalism in Asia*, edited by Yasuhei Taniguchi, Alan Yanovich, and Jan Bohanes (Cambridge University Press, 2007), p. 398.

78. Aaron Cosbey and others, *Clean Energy Investment: Project Synthesis Report* (International Institute for Sustainable Development, July 2008), p. 68.

79. See the International Energy Forum (www.energyforum.gov.sa).

80. See the Joint Oil Data Initiative (www.jodidata.org).

81. Adil Najam and Cutler J. Cleveland, "Energy and Sustainable Development at Global Environmental Summits: An Evolving Agenda," *Environment, Development, and Sustainability* 5 (2003), pp. 117–38.

82. Gleneagles Communiqué on Climate Change, Clean Energy, and Sustainable Development, July 2005 (www.g7.utoronto.ca/summit/2005gleneagles/communique.pdf).

83. World Bank, "An Investment Framework for Clean Energy and Development: A Progress Report," DC2006-0012, September 5, 2006 (http://siteresources.worldbank.org/DEVCOMMINT/Documentation/21046509/DC2006-0012(E)-CleanEnergy.pdf).

84. Ibid., p. 9.

85. Ibid., p. 10.

86. Friends of the Earth International, "How the World Bank's Energy Framework Sells the Climate and Poor People Short: A Civil Society Response to the World Bank's Framework for Clean Energy and Development," September 2006 (www.foei.org/en/publications/pdfs/wbenergyreport.pdf/view).

87. See, for example, World Bank, *The World Bank's Role in the Electric Power Sector: Policies for Effective Institutional, Regulatory, and Financial Reform* (Washington, 1993); World Bank, *Energy Efficiency and Conservation in the Developing World* (Washington, 1993); World Bank, *Fuel for Thought: An Environmental Strategy for the Energy Sector* (Washington, 2000).

88. Friends of the Earth International, "How the World Bank's Energy Framework Sells the Climate and Poor People Short," p. 3. For detailed critique of the World Bank's performance on energy and development, see Friends of the Earth, *Power Failure: How the World Bank Is Failing to Adequately Finance Renewable Energy for Development* (Washington: October 2005) (www.foe.co.uk/resource/reports/power_failure_how_the_worl.pdf [April 20, 2009])

89. See "The EU Energy Initiative: A Brief Introduction" (http://ec.europa.eu/development/body/theme/energy/initiative/index_en.htm).

90. See the Global Village Energy Partnership (www.gvep.org).

91. See REEEP (www.reeeep.org).

92. See REN21 (www.ren21.net).

93. See the Asia Pacific Partnership (www.asiapacificpartnership.org).

94. Terra Daily, "Chirac Calls for Beefed-Up UN Environment Agency," February 3, 2007 (www.terradaily.com/2006/070203120425.1a1tbbiv.html [April 14, 2009]).

The timing of the announcement—at a conference held simultaneously with the release of the latest IPCC report—left little doubt that climate change was intended to be the focal point for the revamped agency's work.

95. Colin I. Bradford Jr. and Johannes F. Linn, "Global Governance Reform: Conclusions and Implications," in *Global Governance Reform: Breaking the Stalemate,* edited by Colin I. Bradford Jr. and Johannes F. Linn (Brookings, 2007), pp. 115–16.

96. Richard Scott, *IEA: The First Twenty Years: Origins and Structures,* vol. I (Paris: OECD/IEA, 1994), pp. 184–200.

97. The outline of Barack Obama's energy plan can be read at http://my.barackobama.com/page/content/newenergy.

Climate Change

CHAPTER EIGHT

Features of Climate-Smart Metropolitan Economies

MARILYN A. BROWN, FRANK SOUTHWORTH, and ANDREA SARZYNSKI

Meeting the climate challenge requires the leadership of metropolitan America. The 100 largest metropolitan areas in the United States, which comprise two-thirds of the nation's population and account for nearly three-quarters of its economic activity, are responsible for much of the nation's greenhouse gas (GHG) emissions.[1] At the same time, metropolitan America is the traditional locus of technological, entrepreneurial, and policy innovations. Its access to capital and a highly trained workforce has enabled it to play a pivotal role in expanding U.S. business opportunities while solving environmental challenges. With supportive federal policies, metropolitan areas can provide the climate-smart leadership required to meet the nation's targets and timetables for avoiding the buildup of dangerous levels of atmospheric greenhouse gases.

Many metropolitan actors already are at the forefront of state and national action on the climate. For example, more than 1,000 mayors, representing almost 30 percent of the U.S. population, have signed the U.S. Mayor's Climate Protection Agreement.[2] However, the lack of adequate data on emissions and of comparative analysis makes it difficult to confirm or refute best practices and policies. To help provide benchmarks and expand understanding of carbon emissions, this chapter quantifies highway transportation and residential carbon emissions of the 100 largest U.S. metropolitan areas in 2000 and 2005.[3] The carbon emissions from transportation and residential sources—some of the most consumer-

FIGURE 8-1. Greenhouse Gas Emissions in the United States, 2005

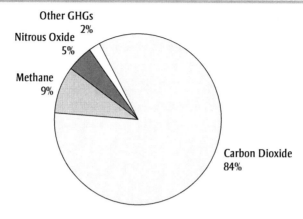

Source: EPA, "Inventory of U.S. GHG Emissions and Sinks: 1990–2006," 2007, table 2-1.

dominant sources of greenhouse gas emissions—provide a foundation for identifying energy pricing, land use, and other policy interventions that could reduce the impact of the U.S. economy on energy consumption and the climate.[4]

The Climate Challenge

Carbon dioxide, which accounted for 84 percent of U.S. greenhouse gas emissions in 2005, is one of the most important contributors to climate change (see figure 8-1). The vast majority of anthropogenic carbon dioxide is released when carbon-based fuels, such as coal and oil, are burned for energy.[5] (Here, the terms "carbon emissions" and "carbon footprint" both indicate emissions of carbon dioxide.) Residential and commercial buildings account for 39 percent of the carbon emissions in the United States. Transportation accounts for one-third of U.S. emissions, and industry is responsible for 28 percent (figure 8-2). An effective climate strategy must focus on reducing carbon emissions across all sectors.

Carbon emissions in the United States have increased by almost 1 percent a year since 1980.[6] Emissions from the residential, commercial, and transportation sectors each have increased by more than 25 percent during the past twenty-five years.[7] Industrial emissions, however, have declined during the same period as the country has moved away from energy-intensive manufacturing and toward a service and knowledge

FIGURE 8-2. CO$_2$ Emissions in the United States by Sector, 2005

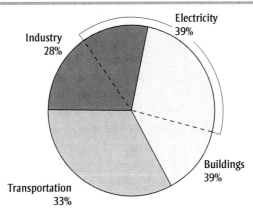

Source: EIA, "Annual Energy Outlook 2007," table A18.

economy. Since much of what the United States once manufactured is now being imported from China, India, and other countries, standard accounts of U.S. greenhouse gas emissions exclude much of what the nation is actually "responsible" for, to use the terminology of Louis Lebel and colleagues.[8]

As a result, consumers increasingly are driving domestic energy consumption and carbon emissions. Residential and commercial buildings and road transportation are expected to dominate the future growth of energy demand and carbon emissions. Total U.S. carbon emissions are projected to grow by 16 percent between 2006 and 2030, making reductions all the more urgent to avoid the worst potential effects of a warming planet.[9] Four factors determine carbon emissions:

—population
—economic output
—energy intensity of the economy
—carbon intensity of the energy system.[10]

Shrinking the nation's carbon footprint, while allowing for population and economic growth, requires a strategic focus on reducing both energy and carbon intensities. To do so requires reducing the amount of energy needed to power the economy and/or reducing U.S. reliance on fuels that emit high levels of carbon, such as coal and petroleum. Reductions can be made in each sector as well as through multisector approaches.

Reductions of the magnitude needed to curtail global climate change—often estimated to be on the order of magnitude of 50 to 80 percent below current emission levels by mid-century—will not be easy. Energy intensity is much higher in the United States than in many other developed countries, resulting in a national footprint of 5.5 metric tons of carbon per capita; the global average, in contrast, is only approximately 1.2 metric tons per person.[11] Despite recent improvements, U.S. energy intensity is high relative to the world average and to the energy intensity of many other developed nations; Japan, for instance, produces a dollar of GDP with less than half of the energy required in the United States.[12] Although China overtook the United States and Europe in 2006 to become the world's largest carbon emitter, the United States will likely remain one of the most carbon- and energy-intensive nations on Earth well into the future.[13]

Transportation

Transportation is responsible for one-third of the nation's carbon footprint. Highway transport accounts for 80 percent of the total, dominated by automobiles (30 percent), light-duty trucks (27 percent), and freight transport (20 percent). Air- and water-based transport is responsible for a majority of the remainder. The transportation sector also has the fastest-growing footprint of the major energy-using sectors. Between 1991 and 2006, transportation accounted for nearly one-half of the growth in U.S. carbon emissions.[14] Trends in highway transport, which contributes the most to transportation emissions, deserve special attention.

Suburbanization and rising wealth following World War II dramatically transformed U.S. living and driving patterns. The country saw a ubiquitous increase not only in daily travel distances but also in the frequency with which households used their vehicles to get to work, to shop, and to carry out a variety of personal activities. Between 1970 and 2005, the average annual vehicle miles traveled (VMT) per household increased almost 50 percent, from 16,400 to 24,300.[15] At the same time, vehicle ownership per household increased even as average household size fell.[16] The increase in the annual rate of commercial truck travel (3.7 percent) was even more than the increase in the rate of passenger travel (2.8 percent).[17] That increase in travel is responsible for worsening traffic congestion, wasted fuel, and rising carbon emissions.[18]

Despite significantly improved automotive engine technologies, miles per gallon (mpg) gains have leveled off since the mid-1980s, in part due to a consumer preference for larger, more powerful vehicles.[19] While significant fuel has been saved by advances in motor technology, most gasoline-fueled vehicles on the road today use only 15 percent of the fuel's energy to move the vehicle down the road and to power accessories such as air conditioning. The rest is lost to engine inefficiencies and idling.[20]

The U.S. transportation sector is powered primarily by gasoline, followed by diesel, which together accounted for 98 percent of U.S. vehicle fuel consumption in 2005. On a "well-to-wheels" basis, diesel is about 15 percent less carbon intensive than gasoline.[21] Thus, greater use of diesel technologies in the U.S. vehicle fleet would improve fuel efficiency and reduce carbon emissions. Improvements in fuels and technology also have the potential to reduce carbon emissions from the transportation sector substantially. Promising developments are taking place in hybrid electric and cellulosic biofuel vehicle technologies. Cellulosic ethanol and biodiesel may prove to be important low-carbon alternatives to gasoline and diesel.[22] For example, replacing one-quarter of the gasoline projected to be used with cellulosic ethanol—a replacement rate viewed as achievable within twenty-five years—could cut carbon emissions by 15 to 20 percent.[23] Another promising alternative is hybrid electric systems that are recharged in off-peak hours by low-carbon electricity. Metropolitan areas are especially well suited to low-carbon options because the capital investment needed to establish new refueling infrastructures is more economically feasible in high-density environments.

Under the Energy Independence and Security Act (EISA) of December 2007, automakers are required from 2011 on to increase the fuel economy of passenger vehicles by 40 percent, to a fleet average of 35 mpg by 2020.[24] In addition, the federal government is directed to study and work toward "maximum feasible" fuel economy standards for small (8,500–10,000 pound) "work" trucks as well as medium-size and large commercial trucks. The production in recent years of a number of higher-mpg automobiles suggests that significant increases in vehicle and truck fuel economy appear feasible as well as justifiable. Increases could be achieved through the introduction of higher-mpg conventional gasoline vehicles as well as diesel-fuel vehicles and through the rapidly growing market for gasoline-electric hybrids, which can attain on the road fuel efficiencies well above the current 35-mpg national fuel economy standard set for 2020. [25]

After the reductions mandated by EISA are accounted for, transportation energy use is projected to grow by 0.4 percent annually.[26] Such an increase in energy use could drive up transportation carbon emissions 10.3 percent between 2006 and 2030.[27] During the same period, crude oil imports are forecast to rise from 66 to 71 percent of total supply, increasing U.S. vulnerability to disruptions in the supply and price of petroleum. In the transportation sector in particular, energy and climate challenges are intertwined with energy security concerns.[28]

Buildings

Through the energy that they use, buildings are responsible for 39 percent of U.S. carbon emissions. Single-family homes, apartments, manufactured housing, and other residential buildings account for slightly more than one-half of emissions, with commercial buildings (offices, businesses, hospitals, hotels, and so forth) responsible for the remainder. In the United States, more than one-half of residential energy consumed comes from electricity: 65 percent in 2000 and 68 percent in 2005.[29] Households use electricity for cooling (and some heating), for lighting, and increasingly for televisions, computers, and other household electronics.[30] More than one-half of the electricity in the United States is generated by coal-fired central power plants that have operated at about 35 percent efficiency for more than a half century. Almost two-thirds of the energy embodied in coal is lost through the release of low-temperature waste heat, either at the power plant or along its route to the end user.[31] Depending on how the electricity is ultimately used, as much as 98 percent of the energy in the coal used to produce electricity can be lost as waste heat.[32]

The balance of U.S. residential energy consists of direct fuel consumption. Natural gas is the most common source of heating in buildings and also is used for cooking and heating water. On an energy basis, natural gas has the lowest carbon intensity of fossil fuels.[33] Other low-carbon energy options exist but are not widely used in buildings, including solar photovoltaics, solar lighting, and solar water heating, which are virtually carbon-free, and geothermal heat pumps, which are a low-carbon source of heating and cooling.

The United States has made remarkable progress in reducing the energy use and carbon intensity of its building stock and operations. Those improvements are largely the result of advances in the energy efficiency of U.S. buildings following the 1973–74 OPEC oil embargo, which were motivated in part by the significant proportion of electricity generated from petroleum fuels and the greater reliance on fuel oil for home heating at that time. Since 1972, building energy use overall has increased at less than half the rate of growth of the nation's gross domestic product (GDP) and residential energy use per household has declined.[34] At the same time, homes have grown larger and households use a broader range of equipment, especially air conditioning during the summer and electronic devices throughout the year.

Despite such impressive efficiency gains, the total energy used in buildings almost doubled between 1970 and 2005, and the nation can expect to see building energy consumption increase by 0.8 percent a year through 2030.[35] Because of the predominance of the use of electricity in the building sector and the anticipated expansion of the nation's building stock to accommodate population growth, carbon emissions from the built environment are expected to grow rapidly. While that new growth is occurring, most of the current stock of buildings will continue to be occupied, although much of it will have been redeveloped, which presents the opportunity to upgrade current buildings to eco-friendly features as they are developed and become more economically feasible and widely available.

Development Patterns

The spatial arrangement of buildings and transportation infrastructure in communities and urban systems can play a role in carbon reduction. Urban form links the energy consumed in different building designs, densities, and land-use configurations to the energy required to support daily travel, provide freight pickups and deliveries, and support a rapidly growing number of on-the-job service trips. The carbon-reduction benefits realized from building more spatially compact, mixed-use developments that also have access to rapid transit include numerous complementary effects:

—reduced residential heating and cooling costs due to smaller homes and shared walls in multi-unit dwellings

—use of district heating and cooling systems using centralized boilers that enable the cogeneration of electricity from waste heat

—lower energy losses along electricity distribution lines from locating power generators (for example, gas turbines or building-integrated solar photovoltaics) closer to electricity users

—shorter freight and personal trips, as well as the use of public transit, walking, and cycling for those trips

—reduced municipal infrastructure requirements, including less need for local street construction and shorter electric, communication, water, and sewage lines

—use of micro grids to meet local electricity requirements with highly efficient distributed power generation

—reuse and retrofitting of existing structures.

Some studies have quantified the role of compact development in carbon reductions. For instance, the number of dwellings per acre is directly related to GHG emissions. With shared walls and generally smaller square footage, households in buildings that include five or more units consume only 38 percent of the energy of households in single-family homes.[36] At a suburban density of four homes per acre, carbon dioxide emissions per household were found to be 25 percent higher than in an urban neighborhood with twenty homes per acre.[37]

Studies also show that household vehicle miles traveled vary with residential density and access to public transit.[38] Higher housing and job densities, mixed land uses, and a balanced jobs–housing ratio within an area are associated with shorter trips and lower automobile ownership and use.[39] In comparing two households that are similar in all respects except residential density, the household in a neighborhood with 1,000 fewer housing units per square mile drives almost 1,200 miles more and consumes 65 more gallons of fuel a year than its peer household in a higher-density neighborhood.[40]

Less is known about how household behavior may change in response to changes in density or the concentration of housing or jobs. A recent simulation estimates that shifting 60 to 90 percent of new growth to development that is more compact than current developments, which increase urban sprawl, would reduce VMT by 30 percent and cut U.S. carbon dioxide emissions from transportation by 7 to 10 percent by 2050.[41] That effect is comparable to what might occur if fuel prices doubled.[42] However, it may be unrealistic to expect 60 to 90 percent of new growth to take place in compact developments, suggesting that compact development might play a secondary role to advances in efficiency, technology, and fuels. Other efficiency studies project even greater and more rapid GHG reductions from compact development, with savings of 10 percent of the U.S. 2001 level of GHGs possible within as few as ten years— although, again, such predictions may be optimistic.[43]

A Partial Carbon Footprint of Metropolitan America

Metropolitan areas form the backbone of the U.S. economy. Before researchers can evaluate the impact of existing carbon reduction efforts and of proposed policy changes, the nation needs a consistent set of emissions data for multiple periods and of a scale that can be tied to the activities, land uses, and infrastructure of metropolitan areas.

Our study begins to fill that need by producing comparable partial carbon footprints for the 100 largest metropolitan areas in 2000 and 2005. These footprints are based on national databases for passenger and freight highway transportation and for energy consumption in residential buildings; they do not include emissions from commercial buildings, industry, or non-highway transportation (that is, air, water, transit, or rail transportation).[44] The footprints also measure only carbon dioxide emissions from use of fossil sources of energy; the impact of urban development on deforestation and other possibly significant impacts on the atmospheric GHG balance are not considered.

Analysis of the partial carbon footprints of the top 100 largest metropolitan regions in the United States reveals five major findings:

Large metropolitan areas offer greater energy and carbon efficiency than non-metropolitan areas. Despite housing two-thirds of the nation's population and accounting for three-quarters of its economic activity, in 2005 the nation's 100 largest metropolitan areas emitted just 56 percent of U.S. carbon emissions from highway transportation and residential buildings. Consequently, residents of metropolitan areas have smaller partial carbon footprints than the average American. The average metropolitan area resident's partial carbon footprint (2.24 metric tons) in 2005 was only 86 percent of the average American's partial footprint (2.60 metric tons). The difference is due primarily to less car travel and less use of residential electricity than to less freight travel and less use of residential fuels.

Between 2000 and 2005, carbon emissions increased more slowly in metropolitan areas of the United States than in the rest of the country. Carbon emissions from highway transport and residences in major metropolitan areas increased 7.5 percent from 2000 to 2005, slightly less than the national increase of 9.1 percent. The population of the 100 largest metropolitan areas, on the other hand, grew by only 6.3 percent.

As a result, the average per-capita footprint of the 100 metropolitan areas grew by 1.1 percent during the five-year period, while the U.S. partial carbon footprint increased twice as rapidly (by 2.2 percent) during the same timeframe. While seventy-nine metropolitan areas saw overall growth in their carbon emissions from highway transport and residential energy use from 2000 to 2005, only fifty-three metropolitan areas increased their footprints on a per-capita basis. Another twenty-one metropolitan areas saw their carbon emissions from highway transport and residential energy use decline.

In the 100 largest metropolitan areas and the nation overall, carbon emissions grew faster from 2000 to 2005 for auto transport and residential electricity use than for freight travel and residential fuels. Trenton, New Jersey, and Chattanooga, Tennessee, saw the most growth in both total carbon emissions and per-capita footprint.[45] Conversely, Youngstown, Ohio, and Grand Rapids, Michigan, each saw its carbon footprint decline by 14 percent during the five-year period—the largest declines in the 100 metropolitan areas. Both of these urban areas suffered serious losses of economic activity over that period, which undoubtedly contributed to their shrinking carbon signatures. In contrast, Riverside and Bakersfield, California, and El Paso, Texas, reduced their per-capita footprints by more than 10 percent despite increasing their total emissions.

Reversing the rising trend in emissions—which many climate scientists warn must happen to mitigate the effects of climate change—poses a distinct challenge for many metropolitan areas and the nation as a whole. According to the data for 2000 and 2005, metropolitan America is constraining the growth of its carbon footprints better than nonmetropolitan areas.

Per-capita carbon emissions vary substantially by metropolitan area. In 2005, per-capita carbon emissions were highest in Lexington, Kentucky, and lowest in Honolulu. The average resident in Lexington emitted 2.5 times more carbon from transport and residential energy use in 2005 than the average resident in Honolulu (3.46 metric tons and 1.36 metric tons respectively). While readers might immediately note the different climatic conditions of the two urban areas—Lexington has a combination of winter heating and summer cooling "loads," and Honolulu has a Mediterranean climate that requires much less space conditioning—the factors affecting the wide gap between their carbon footprints are in fact much more complex.

Variations are even more striking after adjusting for a metro area's economic output—or gross metropolitan product (GMP)—which is an indicator of *carbon intensity*. In this case, the carbon footprints range from a high of 97.6 million metric tons of carbon per dollar of GMP in Youngstown, Ohio, to a low of 22.5 million metric tons per dollar of GMP in San Jose, California—more than a four-fold difference. While the

FIGURE 8-3. Map of Per Capita Carbon Footprints, 2005

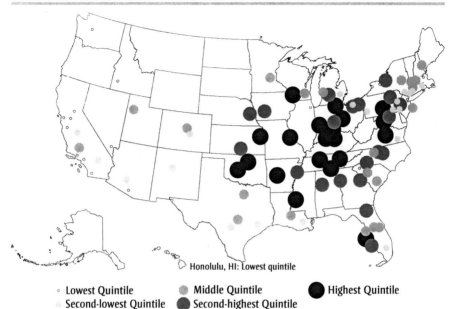

Honolulu, HI: Lowest quintile

○ Lowest Quintile   ⦿ Middle Quintile   ● Highest Quintile
Second-lowest Quintile   ● Second-highest Quintile

two extremes compare a traditional Rust Belt area with a Silicon Valley information economy, keep in mind that the carbon footprints in our study are measured only by emissions from residential and transportation activities. Thus, they do not reflect what undoubtedly would be an even more pronounced difference between the two areas if carbon emissions from their industrial activities were included. In other contrasts, residents in Nashville and St. Louis (Missouri) emitted twice as much carbon from transport and residential uses, on average, than did residents in San Francisco and Seattle.

Regional variation in carbon emissions is apparent. Most notably, the Mississippi River roughly divides the country into high emitters and low emitters (see figure 8-3). In 2005, all but one of the ten largest per-capita emitters—Oklahoma City being the exception—were located east of the Mississippi. On the other hand, all but one of the ten lowest per-capita emitters—New York being the exception—were located west of the Mississippi. California alone was home to six of the twenty lowest per-capita emitters.

FIGURE 8-4. Map of Transportation Per Capita Carbon Footprints, 2005

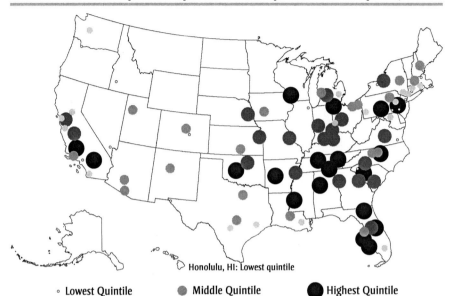

Honolulu, HI: Lowest quintile

○ Lowest Quintile   ◉ Middle Quintile   ● Highest Quintile
◌ Second-lowest Quintile ● Second-highest Quintile

Eight of the 20 highest emitters were located east of the Mississippi River and south of Ohio (including Cincinnati, which straddles the Ohio-Kentucky border).

The West is the only region that reduced its partial carbon footprint between 2000 and 2005; the Midwest, Northeast, and South all increased their per-capita carbon emissions. Reflecting the rapid growth and decentralization of many Southern cities, the carbon footprints of metropolitan areas in the South grew more rapidly than in any other region. The South also has the dubious distinction of having the largest carbon footprints from both transport and residential uses of any region in both 2000 and 2005. Fourteen of the twenty metropolitan areas with the largest transportation footprints are in the census-defined South, and half of the twenty with the largest residential footprints are in the South (see figures 8-4 and 8-5). Despite such geographic clustering, only five metropolitan areas appear in the top-twenty list for both transportation and residential energy.

Development patterns and rail transit influence carbon emissions.[46] Population density (that is, the number of people per acre of developable land),

FIGURE 8-5. Map of Residential Per Capita Carbon Footprints, 2005

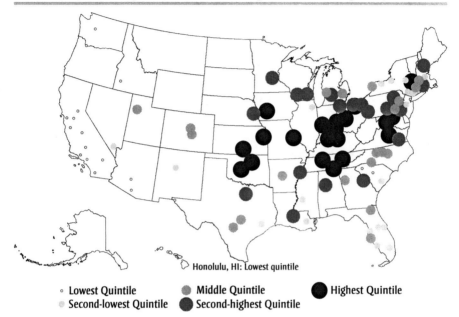

Honolulu, HI: Lowest quintile

○ Lowest Quintile ◉ Middle Quintile ● Highest Quintile
◌ Second-lowest Quintile ● Second-highest Quintile

concentration of development (referring to the evenness of population density), and rail transit (based on a threshold number of miles of rail transit lines) all tend to be higher in the lowest-emitting metropolitan areas.[47] Much of what appears as regional variation may actually be due to these spatial factors, as many of the older, denser cities in the Northeast, Midwest, and California (for example, Boston, New York, Chicago, and San Francisco) are all low emitters.

Generally, knowing a metropolitan area's overall density helps predict its carbon emissions.[48] Dense metropolitan areas such as New York, Los Angeles, and San Francisco stand out for having the smallest transportation and residential footprints. Alternatively, low-density metropolitan areas such as Lexington, Nashville, and Oklahoma City are prominent among the ten largest per-capita emitters.

The benefits of density are not necessarily unique to metropolitan areas. The 100 largest metropolitan areas appear to perform better than the rest of the country because of their overall density. However, large metropolitan areas have a patchwork of higher- and lower-density areas—density is not uniform across the entire metropolitan area. Therefore, whether in metropolitan areas or small towns, higher-density developments have smaller transportation and residential carbon footprints. That pattern is confirmed by examining population and employment concentration measures, which reflect clustering at the ZIP code scale.[49] Compact development also generates other benefits for its residents, such as the health, safety, and social benefits offered by walkable communities.[50]

Many metropolitan areas with small per-capita footprints also have sizable rail transit ridership. New York, San Francisco, Boston, and Chicago have some of the highest annual rates of rail ridership in the nation, ranging from 296 to 757 miles per capita, and carbon footprints ranging from 1.5 to 2.0 tons of carbon per capita—much lower than the average of 2.2 tons for all 100 metropolitan areas. Looking just at carbon footprints from highway transportation highlights a cluster of low emitters located along the Washington to Boston corridor. In addition to benefiting from rail transit, these cities also tend to have the high population densities characteristic of older cities of the Northeast. There are exceptions to the rail-footprint connection. Washington, D.C., Baltimore, and Atlanta, for example, all have high rail transit ridership but also have substantially larger-than-average carbon footprints, underscoring the multidimensional nature of carbon footprints.

Finally, freight traffic poses a problem for metropolitan areas trying to shrink their carbon footprints. For example, in 2005, Bakersfield, California, had the smallest residential footprint in the sample (at 0.35 metric tons per capita) but the largest transportation footprint (at 2.2 metric tons), largely because of its freight traffic. Jacksonville and Sarasota, Florida, and Riverside, California, are similar, having the sixth-, seventh-, and ninth-largest transportation footprints, combined with lower-than-average residential carbon footprints. All three metropolitan areas have or are near port cities with sizable freight traffic. They also report significant miles of travel by combination trucks, which typically involve low-efficiency trips that either start or end outside the metropolitan area's boundaries (contributing to what Louis Lebel and colleagues call the "logistics" part of a city's carbon footprint).[51]

Other factors, such as local climate, the fuels used to generate electricity, and electricity prices also influence carbon footprint. Some areas may perform well on transportation but have large residential footprints. Cleveland, Ohio; Springfield, Massachusetts; and Providence, Rhode Island, fit that model. They fall among the twenty-five lowest emitters for highway transportation but are in the top twenty-five for residential emissions. These metropolitan areas have high emissions from residential fuels, as do many other northeastern and midwestern metropolitan areas.

Climate unmistakably plays a role in residential footprints. Many areas in the Northeast, for instance, have large residential footprints because of their heavier reliance on carbon-intensive home-heating fuels such as fuel oil. Warm areas in the South often have large residential footprints because of their heavy reliance on carbon-intensive air conditioning. High-emitting metropolitan areas concentrate throughout the mid-latitude states of the eastern United States, where combined cooling and heating requirements are substantial. In contrast, the ten metropolitan areas with the smallest per-capita residential footprints are all located along the Pacific coast, with its milder climate.

The fuel mix used to generate electricity matters in residential footprints. For instance, in 2005, the Washington, D.C., metropolitan area's residential electricity footprint was ten times larger than Seattle's.[52] The mix of fuels used to generate electricity in the nation's capital includes high-carbon sources like coal, while Seattle draws its energy primarily from essentially carbon-free hydropower. A high-carbon fuel mix significantly penalizes the Ohio Valley and Appalachian regions, which rely heavily on power produced by coal. The investor-owned utilities in some states, such as California, no longer purchase electricity from coal-powered plants, resulting in lower residential carbon footprints.

Electricity prices also appear to influence the residential footprint.[53] Each of the ten metropolitan areas with the lowest per-capita electricity footprints in 2005 hailed from states with higher-than-average prices, including California, New York, Michigan, and Hawaii. On the other hand, many southeastern metropolitan areas with high electricity consumption have had historically low electricity rates.

Summary

Our findings show that large metropolitan areas offer greater energy and carbon efficiency than nonmetropolitan areas and that metropolitan area development patterns show promise for reducing carbon emissions. Three pressing challenges, however, remain for metropolitan America. First, between 2000 and 2005, carbon footprints grew faster than the population in the 100 largest metropolitan areas and in the nation at large. Second, many of the fastest-growing metropolitan areas are also the least compact, such as Austin, Texas; Raleigh, North Carolina; and Nashville, Tennessee. Third, some important factors may be largely beyond the control of metropolitan areas, such as local climate. Fortunately, many obstacles can be addressed by policy interventions. In the long run, however, metropolitan America will be hard-pressed to shrink its carbon footprint in the absence of supportive federal policies.

Climate-Smart Policies

Unlike the population of Europe, Japan, and many other developed economies, the U.S. population is expected to continue to expand, rising from 300 million today to 420 million in 2050, according to U.S. Census Bureau projections.[54] As the U.S. population grows, the nation must reduce the energy intensity of its economic system and lower the carbon intensity of its energy consumption. Because such transformations require capital, often they are cost-effective only when capital assets are first being built or when major upgrades, renovations, or system replacements are undertaken. If improved technology is not installed at such times, the carbon-intensive status quo can be locked in for decades.[55]

The current U.S. economic downturn offers a period of reflection and an opportunity to prepare for future demands. Almost $40 billion of the $787 billion appropriation under the American Recovery and Reinvestment Act of 2009 (the "Stimulus Bill") is available to invest in climate-smart infrastructure and facilities. That includes, for example, $3.2 billion to fund a new energy efficiency and conservation block grant (EECBG) program for state, local, and tribal governments to use for energy efficiency and conservation projects. Such resources need to be dedicated to high-payoff investments that will facilitate the country's transition to a low-carbon economy.

Our research suggests that high-payoff investments are likely to come from investing in the nation's metropolitan areas, which provide opportunities for more energy- and carbon-efficient lifestyles. Such investment makes sense because almost all of the nation's built environment and energy infrastructure is concentrated in metropolitan areas and a high percentage of the country's future growth will take place in metropolitan areas.

The existence of low-cost opportunities to create climate-friendly metropolitan environments does not necessarily mean that decisionmakers and consumers will select low-carbon alternatives. Numerous flaws prevent the market from operating efficiently in tackling the climate challenge. Correcting those flaws requires major economy-wide public policies as well as actions at the metropolitan level.

The most important economy-wide market failure is the lack of a price on carbon emissions. Thus, a key remedy involves getting energy prices right by internalizing the climate costs of fossil fuel combustion through carbon taxes or a cap-and-trade system.[56] Carbon pricing is arguably the most efficient policy mechanism to encourage efficient and low-carbon energy choices, but realistically it can be implemented only at the national or international level. More local policies are prone to carbon leakage, spillovers, and free riding.[57] The federal government must also create new programs and policies and expand others to encourage decisionmaking that shrinks the nation's carbon footprint. Such actions include increasing spending on energy research, development, and demonstration (RD&D), developing a national renewable electricity standard, and providing better data and technical assistance to states and localities (see table 8-1).

In addition, five federal initiatives are needed to promote climate-smart development and ensure success in metropolitan America (table 8-1). First, federal transportation policy should place highway and transit funding decisions on an equal footing, which would encourage new *transit-oriented development* and redevelopment of existing urban spaces. That in turn would improve prospects for reducing the nation's transportation footprint through expanded use of public transit and nonmotorized travel.

Second, the federal government should facilitate more energy-efficient freight operations, which concentrate in the nation's largest metropolitan areas, starting with the establishment of more effective *regional freight planning* that considers both intra- and inter-metropolitan freight operations. Opportunities for reducing the freight carbon footprint include use and maintenance of more energy-efficient vehicles, introduction of more energy-efficient intra-urban trucking operations, and development and operation of more energy-efficient freight intermodal terminals. Freight carriers as well as their customers stand to gain financially from more fuel-efficient operations, but they will need to be convinced of the monetary as well as environmental benefits of making the necessary changes. Efforts such as EPA's SmartWay transport program, which informs trucking companies of ways that they can reduce their fuel bills and the associated carbon emissions, offer mechanisms by which the federal government can not only promote but also support (through innovative financing mecha-

TABLE 8-1. Ten Recommendations to Help Correct Flaws in Current Federal Policy

Flaw	Economy-wide policies
Fossil energy is underpriced.	Put a price on carbon to account for the external costs of fossil fuel combustion.
Energy RD&D is underfunded by the federal government.	Increase funding of energy RD&D to increase energy-efficient and low-carbon innovations and accelerate their use.
National standards are lacking.	Establish a national renewable electricity standard to foster low-carbon energy markets in a rational and predictable policy environment.
State utility pricing policies and cost-recovery regulations thwart improvements in energy efficiency and adoption of low-carbon options.	Help states reform their electricity regulations to promote energy efficiency.
Information on local GHG emissions and best practices is inadequate.	Improve collection and dissemination of information on emissions and best practices for states and localities

Flaw	Targeted policies
Federal transportation policy makes more energy-efficient development patterns less viable.	Promote more transportation choices to expand transit-oriented and compact development options.
The federal government defers to state and local governments in regulating land use.	Develop regional freight planning to introduce more energy-efficient freight operations.
The federal government's housing and building code policies do not adequately promote energy efficiency in buildings. Federal incentives for energy-efficient investments are biased in favor of new homes and higher-income households.	Require disclosure of energy costs and "on-bill" financing to stimulate and scale up energy-efficient retrofitting.
Federal transportation policy inhibits energy-efficient development patterns. Mortgage tax policy and lending practices hinder climate-friendly development. The federal government fails to leverage its housing finance programs to stimulate energy-efficient building.	Use federal housing financing to create incentives for location-efficient mortgages and to reform policies that lead to the overconsumption of housing.
All of the above.	Issue a metropolitan challenge to reward metro areas for developing innovative spatial solutions.

nisms) the adoption of greener transportation options in metropolitan areas.[58]

Third, the federal government should make efforts designed to improve housing decisions, such as by requiring greater *disclosure of home energy costs* and *"on-bill" financing* options, which would help to upgrade the energy integrity of the nation's building stock. With such disclosures, buyers can gauge energy costs and how those costs may be influenced by the building's current features. In one of the first examples in the United States, Austin, Texas, passed an ordinance in 2008 that combines a requirement to conduct an energy audit before a home is sold with a voluntary program for implementing cost-effective upgrades; it also sets targets for audits of multifamily units.[59] With on-bill financing, a utility company (or state or federal agency) loans money for the purchase of energy-efficient equipment to consumers, who repay the loans in monthly utility bill payments that are no greater than the monthly energy savings.[60] This financing mechanism provides a way for homeowners to save money in the long term. The effectiveness of this type of program is greatly enhanced by partnering with utilities because they already have an established billing relationship their customers and have access to information about customers' patterns of energy use and payment history.

Fourth, *federal housing financing* should be used to create incentives for energy- and location-efficient housing choices. The federal government has an opportunity to construct market-catalyzing financial products, such as energy-efficient and location-efficient mortgages (EEMs and LEMs). It also should reconsider the mortgage interest deduction, which encourages people to buy more and larger homes on larger lots in less-dense locations.[61] Current mortgage-lending practices encourage homebuyers to "drive until they qualify," by seeking more "affordable" housing farther from the urban core. Homes on the urban fringe become less affordable when energy prices climb, as illustrated when gasoline prices spiked in 2008.[62] Climate-smart housing policies would encourage repopulating the urban core and reducing sprawl while reducing energy consumption.

Finally, the federal government should issue a metropolitan challenge grant to encourage metropolitan areas to shrink their carbon footprints by integrating housing, transportation, and economic development policies. Without such holistic approaches, metro actors will be hard-pressed to develop the place-based transformative policies needed to address climate and energy challenges.

Conclusion

Metropolitan areas and the built environment have for the most part been left out of the discussion of future actions to mitigate global climate change and strengthen energy security. Yet metropolitan areas have provided climate-smart leadership, and they could play a much bigger role in the future. Together, a federal portfolio of metropolitan carbon policies could place a supportive America squarely at the forefront of solutions to the nation's energy and climate challenges.

Notes

1. For a list of areas, see the appendix in Alan Berube, "MetroNation: How U.S. Metropolitan Areas Fuel American Prosperity" (Metropolitan Policy Program, Brookings, 2007).

2. Steve Nicholas, *Summit on America's Climate Choices* (Washington: Institute for Sustainable Communities, March 30, 2009).

3. Metropolitan areas were selected on the basis of their total employment as of 2005.

4. This chapter draws from research conducted for the Brookings Institution. See Marilyn A. Brown, Frank Southworth, and Andrea Sarzynski, *Shrinking the Carbon Footprint of Metropolitan America* (Brookings, May 2008); Marilyn A. Brown, Frank Southworth, and Andrea Sarzynski, "The Geography of Metropolitan Carbon Footprints," *Policy and Society* 27 (2009), pp. 285–304.

5. Fuel combustion produced 94.2 percent of the carbon dioxide emitted in the United States in 2006. Environmental Protection Agency, "Inventory of U.S. GHG Emissions and Sinks: 1990–2006" (2007).

6. Energy Information Administration, "Annual Energy Review" (2007), table 12-1.

7. Ibid.

8. See U.S. Department of Energy, "U.S. Energy Intensity Indicators: Trend Data" (http://intensityindicators.pnl.gov/trend_data.stm); Louis Lebel and others, "Integrating Carbon Management into the Development Strategies of Urbanizing Regions in Asia," *Journal of Industrial Ecology* 11 (2007), pp. 61–81.

9. Energy Information Administration, "Annual Energy Outlook 2008" (2008), table A18; Intergovernmental Panel on Climate Change, "Climate Change 2007: The Physical Science Basis: Summary for Policymakers" (2007).

10. Yoichi Kaya, "Impact of Carbon Dioxide Emission Control on GNP Growth: Interpretation of Proposed Scenarios," paper presented to the IPCC Energy and Industry Subgroup, Response Strategies Working Group, Paris, 1990.

11. The U.S. carbon footprint is derived from the Energy Information Administration, "Annual Energy Outlook 2008," table A18. The global carbon footprint, based on data in Energy Information Administration, *Emissions of Greenhouse Gases Report*, DOE/EIA-0573 (Department of Energy, December 3, 2008), is computed by dividing global emissions (28.1 billion metric tons of carbon dioxide) by the world

population (6.4 billion) and dividing the result by 3.67 to convert emissions in units of carbon dioxide to emissions in units of carbon.

12. Council on Competitiveness, "Competitiveness Index: Where America Stands" (2006).

13. Ibid.

14. Energy Information Administration, "Annual Energy Outlook"; Frank Gallivan and others, "The Role of TDM and Other Transportation Strategies in State Climate Action Plans," *TDM Review* (2007), pp. 10–14.

15. Bureau of Transportation Statistics, "National Transportation Statistics 2007" (2007), table 1-32.

16. See Oak Ridge National Laboratory, "Transportation Energy Data Book," table 8-5 (http://cta.ornl.gov/data/index.shtml); see U.S. Census Bureau, "Families and Living Arrangements," table HH-1 (www.census.gov/population/www/socdemo/hh-fam.html).

17. Bureau of Transportation Statistics, "National Transportation Statistics 2007," table 1-32.

18. David Schrank and Tim Lomax, "The 2007 Urban Mobility Report" (College Station, Tex.: Texas Transportation Institute, 2007).

19. Environmental Protection Agency, "Light-Duty Automotive Technology and Fuel Economy Trends: 1975 through 2007" (2007).

20. See Department of Energy and Environmental Protection Agency, "Advanced Technologies and Energy Efficiency" (www.fueleconomy.gov/feg/atv.shtml).

21. See Energy Information Administration, "Light-Duty Diesel Vehicles: Efficiency and Emissions Attributes and Market Issues," February 2009 (www.eia.doe.gov/oiaf/servicerpt/lightduty/chapter2.html).

22. The Energy Independence and Security Act of 2007 extends and adds to the 2005 Energy Policy Act Renewable Fuels Standard by setting a goal of production of 36 billion gallons of renewable fuel annually by 2022, including 16 billion gallons from cellulosic sources. See "H.R. 6: Energy Independence and Security Act of 2007" (www.govtrack.us/congress/bill.xpd?bill=h110-6).

23. National Commission on Energy Policy, "Ending the Energy Stalemate: A Bipartisan Strategy to Meet America's Energy Challenges" (2004).

24. Fred Sissine, "Energy Independence and Security Act of 2007: A Summary of Major Provisions" (Washington: Congressional Research Service, 2007).

25. See Department of Energy and Environmental Protection Agency, "Choosing a More Efficient Vehicle" (www.fueleconomy.gov/feg/choosing.shtml); Department of Energy, Alternative Fuels and Advanced Vehicles Data Center, "HEV Sales by Model: 1999–2008" (www.afdc.energy.gov/afdc/data/vehicles.html).

26. Energy Information Administration, "Annual Energy Outlook," table A2. Recently released 2008 estimates are substantially reduced from 2007 estimates.

27. Ibid., table A18.

28. David B. Sandalow, *Freedom from Oil: How the Next President Can End the United States' Oil Addiction* (New York: McGraw Hill, 2007).

29. Energy Information Administration, "Annual Energy Review," table 2-1b.

30. Ibid., table A18.

31. Thomas R. Casten and Robert U. Ayres, "Energy Myth Eight: Worldwide Power Systems Are Economically and Environmentally Optimal," in *Energy and*

American Society: Thirteen Myths, edited by Benjamin K. Sovacool and Marilyn A. Brown (New York: Springer, 2007).

32. See National Academy of Sciences, "What You Need to Know About Energy," 2008, p. 8 (www7.nationalacademies.org/energy/energy_booklet_pdf.pdf).

33. Environmental Protection Agency, "Inventory of U.S. Greenhouse Gas Emissions and Sinks: 1990–2001" (2007), appendix B.

34. Marilyn A. Brown, Frank Southworth, and Therese Stovall, "Towards a Climate-Friendly Built Environment" (Washington: Pew Center on Global Climate Change, 2005). See also Energy Information Administration, "Annual Energy Outlook," table A2.

35. Energy Information Administration, "Annual Energy Outlook," table A18.

36. Brown, Southworth, and Stovall, "Towards a Climate-Friendly Built Environment."

37. Patrick Mazza, "Transportation and Global Warming Solutions," *Climate Solutions Issue Briefing* (Seattle, Wash.: Climate Solutions, May 2004), pp. 1–4.

38. John Holtzclaw, "A Vision of Energy Efficiency" (Washington: American Council for an Energy-Efficient Economy, 2004).

39. Mary Jean Bürer, David Goldstein, and John Holtzclaw, "Location Efficiency as the Missing Piece of the Energy Puzzle: How Smart Growth Can Unlock Trillion Dollar Consumer Cost Savings" (Washington: Natural Resources Defense Council, 2004).

40. Thomas F. Golob and David Brownstone, "The Impact of Residential Density on Vehicle Usage and Energy Consumption," February 1, 2005 (http://repositories. cdlib.org/itsirvine/wps/WPS05_01 [April 15, 2009]).

41. Reid Ewing and others, "Growing Cooler: The Evidence on Urban Development and Climate Change" (Washington: Urban Land Institute, 2007).

42. Given a –0.3 long-term elasticity of VMT with respect to fuel price, a doubling of fuel prices would reduce VMT by 30 percent. See Victoria Transport Policy Institute, "Transportation Elasticities: How Prices and Other Factors Affect Travel Behavior" (www.vtpi.org/tdm/tdm11.htm).

43. Bürer, Goldstein, and Holtzclaw, "Location Efficiency as the Missing Piece"; Holtzclaw, "A Vision of Energy Efficiency."

44. Other than by including the primary energy content of electricity by accounting for the energy lost in electricity generation and transmission, we do not use a full life-cycle assessment of carbon emissions. That means, for instance, that we do not include the carbon emissions embodied in the energy required to produce manufactured goods such as cars, trucks, and buildings. We also slightly underestimate the GHG emissions from transportation fuels, which, according to EPA's 2006 *Greenhouse Gas Emissions from the U.S. Transportation Sector 1990–1993* (appendix B), amount to a fuel-cycle multiplier for gasoline of about 1.24 to 1.31 or a multiplier of 1.15 to 1.25 for truck diesel.

45. However, the 2000 daily VMT figures for Chattanooga are identified as being unreliable in "Highway Statistics 2000" (table 72), as noted in footnote 43 of Southworth, Sonnenberg, and Brown, "The Transportation Energy and Carbon Footprints of the 100 Largest Metropolitan Areas," Georgia Institute of Technology, School of Public Policy, May 2008 (www.spp.gatech.edu/faculty/workingpapers.php).

46. This and the following finding are based on a multiple regression analysis using eight variables that predicts per-capita transportation and residential carbon

footprints in 2005. Three variables describe a metro area's urban form: population density, population concentration, and presence of rail transit. Two variables describe weather: cooling degree-days and heating degree-days. One variable is the average electricity price in the metro area's primary state. Two control variables are used for the metro area's population size and its economic productivity (output per person). In combination, the six primary explanatory variables explain nearly half (49 percent) of the variation in per-capita carbon footprints across metro areas. Specifically, per-capita carbon footprints are lower in metro areas that have higher population densities, higher concentrations of population, at least ten miles of rail transit infrastructure, fewer cooling degree-days, fewer heating degree-days, and higher electricity prices. Complete urban form measures were not available for Bridgeport, Connecticut; Palm Bay, Florida; and Honolulu, Hawaii, resulting in a sample size of ninety-seven metro areas for the regression analysis. For more information, see Marilyn A. Brown and Cecelia Logan, "The Residential Energy and Carbon Footprints of the 100 Largest Metropolitan Areas," Georgia Institute of Technology, School of Public Policy, May 2008 (www.spp.gatech.edu/faculty/workingpapers. php); Brown, Southworth, and Sarzynski, "The Geography of Metropolitan Carbon Footprints."

47. Various urban form measures, including density, population concentration, and transit availability, were included in preliminary analyses. Population density was defined as the number of persons per acre of "developable" land, which excludes bodies of water and protected lands such as national and state parks. Although this metric is useful, it is incomplete and does not capture spatial distribution patterns. Population concentration was defined as the degree to which population was distributed equally throughout the metro area, using a delta index. The values range from 0 to 1; higher values indicate less clustering and more even distribution of population. For more information, see Frank Southworth, Anthon Sonnenberg, and Marilyn A. Brown, "The Transportation Energy and Carbon Footprints of the 100 Largest Metropolitan Areas."

48. Brown, Southworth, and Sarzynski, "The Geography of Metropolitan Carbon Footprints."

49. For more information, see Southworth, Sonnenberg, and Brown, "The Transportation Energy and Carbon Footprints of the 100 Largest Metropolitan Areas"; Brown, Southworth, and Sarzynski, "The Geography of Metropolitan Carbon Footprints."

50. For an in-depth discussion of the multidimensional concept of compact development, see Ewing and others, "Growing Cooler."

51. Lebel and others, "Integrating Carbon Management into the Development Strategies of Urbanizing Regions in Asia."

52. The carbon content of a metropolitan area's electricity consumption was assumed to be the same as the generation mix of the state in which the area's central city is located. Thus, for the nation's capital, we used the carbon dioxide per megawatt-hour of electricity generated in the District of Columbia, which is one of the highest in the nation because of its reliance on coal. Residents of the Washington metropolitan area actually draw their electricity from a wider region that includes Maryland, Virginia, and possibly other states. If we assumed that Washington's electricity was derived from a lower-carbon source, the area's carbon footprint would be proportionately smaller. Our sensitivity analysis showed that Washington was one of just a few

metropolitan areas whose carbon footprint was subject to significant deviations based on plausible alternative assumptions.

53. Brown, Southworth, and Sarzynski, "The Geography of Metropolitan Carbon Footprints."

54. See Census Bureau, "Projected Population of the United States, by Race and Hispanic Origin: 2000 to 2050," 2004 (www.census.gov/population/www/projections/usinterimproj/natprojtab01a.pdf).

55. Marilyn A. Brown and others, "Carbon Lock-In: Barriers to Deploying Climate Change Mitigation Technologies" (Oak Ridge National Laboratory, 2007).

56. For excellent coverage of potential carbon pricing policies, see "Evaluating the Role of Prices and R&D in Reducing Carbon Dioxide Emissions," Congressional Budget Office, 2006; and Robert N. Stavins, "A U.S. Cap-and-Trade System to Address Global Climate Change" (Brookings, 2007).

57. Benjamin K. Sovacool and Marilyn A. Brown, "Scaling the Policy Response to Climate Change," Policy and Society 27 (2009), pp. 317–28.

58. See Environmental Protection Agency, "Smartway Transport" (www.epa.gov/smartway/transport/index.htm).

59. See City of Austin, Texas, "Ordinance No. 20081106-047," 2008 (www.cityofaustin.org/edims/document.cfm?id=123737).

60. For more information on this emerging public policy, see Marilyn A. Brown and others, "Making Homes Part of the Climate Solution: Policy Options to Promote Energy Efficiency," TM-2009/104, draft (Oak Ridge National Laboratory, April 2009); Matthew Brown, "Brief 3: Paying for Energy Upgrades through Utility Bills" (Washington: Alliance to Save Energy, 2009); Joel Rogers, "Seizing the Opportunity (for Climate, Jobs, and Equity) in Building Energy Efficiency," Center on Wisconsin Strategy, Madison, unpublished manuscript, November 2007.

61. See, for instance, William Gale, Jonathan Gruber, and Seth Stephens-Davidowitz, "Encouraging Homeownership through the Tax Code" (Washington: Urban Institute–Brookings Institution Tax Policy Center, 2007); Joseph Gyourko and Todd Sinai, "The Spatial Distribution of Housing-Related Tax Benefits in the United States" (Cambridge, Mass.: National Bureau of Economic Research, 2001); Richard Voith and Joseph Gyourko, "Capitalization of Federal Taxes, the Relative Price of Housing, and Urban Form: Density and Sorting Effects," Regional Science and Urban Economics 32, no. 6 (2002), pp. 673–90.

62. Christopher B. Leinberger, "The Next Slum?" Atlantic Monthly, March 2008, pp. 70–75.

CHAPTER NINE

Understanding the Interaction between Energy Security and Climate Change Policy

JASON BORDOFF, MANASI DESHPANDE, and PASCAL NOEL

The related topics of energy security and climate change have risen rapidly to the forefront of the U.S. policy agenda. High gas prices topped Americans' list of economic concerns in early and mid-2008, when prices hit an all-time high. Nine of ten Americans said that they expected energy prices to cause them financial hardship in the near term,[1] and even after energy prices fell in late 2008, surveys found that the question of energy prices remained a high concern.[2] At the same time, the ongoing conflict in Iraq and concerns about conflict and instability from Nigeria to Iran to Russia are constant reminders of the vulnerability of world energy supplies. Meanwhile, climate change has become a leading issue in the minds of voters and lawmakers. Vice President Al Gore's documentary on the subject won an Academy Award, and Gore and the Intergovernmental Panel on Climate Change (IPCC) also received the Nobel Peace Prize for their work. Cap-and-trade legislation introduced in the

When this chapter was drafted, in late 2008, the authors were policy director, senior research assistant, and research analyst, respectively, at the Brookings Institution's Hamilton Project. The authors wish to thank Charles Ebinger, Jonathan Elkind, Douglas Elmendorf, Ben Harris, Valerie Karplus, Sarah Ladislaw, Bryan Mignone, Adele Morris, Charles Schultze, and Timothy Taylor for valuable comments and suggestions. Lisa Xu and Julie Anderson provided helpful research assistance. The authors are especially grateful to Matthew Johnson, Evie Zambetakis, and the rest of the Brookings team, who were extremely helpful in finalizing this chapter.

summer of 2008 garnered the support of fifty-four U.S. senators,[3] and both 2008 presidential candidates supported a cap-and-trade measure. In early 2009, the focus continued: the U.S. House of Representatives began consideration of a new cap-and-trade proposal, which passed out of the Energy and Commerce Committee in late May; in addition, some states are taking even greater measures on their own—the northeastern states, for example, created a cap-and-trade program that was set to begin auctioning allowances in September.

The widespread discussion about the need for a better "climate and energy" policy, however, often obscures specific underlying concerns that may be in tension. One worry is that if energy prices are too low, people will consume too much and contribute to climate change. Increased consumption also creates greater dependence on fossil fuels largely produced overseas, often in hostile regions. Yet another worry is that if energy prices are too high, the U.S. economy will suffer and individual households will be adversely impacted. If energy prices are too volatile and dependent on uncertain sources of supply, the economy and individual households will have even greater difficulty managing unexpected shocks.

In short, three related but distinct problems are often misleadingly conflated in discussions about energy policy: climate change, energy security, and energy affordability. To some extent, these issues overlap, but in many underappreciated ways they may conflict. Developing appropriate long-term solutions to these problems requires a precise understanding of their nature and the risks that they pose. Part of the difficulty stems from a lack of agreement over how to define the problems in the first place, particularly for energy security. After defining the challenges of climate change, energy security, and energy affordability more precisely, this chapter explores the interactions between them and lays out appropriate policy responses to each, taking care to minimize the extent to which progress on one gets made at the expense of another.

Defining the Problems

The problems of climate change, energy security, and energy affordability are each often defined in different ways, which exacerbates the difficulty of trying to develop policy responses and understand the interactions between policies. We therefore begin by defining the problems. At a high level, the problems of climate change and energy security are about negative externalities. The climate change, macroeconomic, and national security costs that result from fossil fuel consumption are not directly borne by consumers; they are social costs that economists call negative externalities, and government can help reduce them by making consumers bear them directly. On the other hand, energy affordability is an equity issue. The United States as a society values distributional fairness, and when price increases cause disproportionate hardship for lower-income people, the government should take commonsense steps to address the issue.

Climate change, energy security, and energy affordability problems are discussed in turn below. Because the problem of energy security has an especially wide range of definitions, we focus on it in some detail.

Climate Change

The problem of climate change is clear and accepted by the vast majority of scientists. In the past century, the Earth's average annual surface temperature rose 0.7 degrees Celsius. There is little doubt that humans have contributed to this warming, particularly by burning fossil fuels such as coal and oil. The Intergovernmental Panel on Climate Change asserts with *"very high confidence* [emphasis in original] that the global averaged net effect of human activities since 1750 has been one of warming." Using the best estimate from a range of climate models and future emissions scenarios, the IPCC projects that if emissions continue on their present course, global temperatures will rise another 1.8 to 4.0 degrees Celsius by 2100 and possibly even higher.[4] Such a temperature change may trigger massive climatic shifts, including rising sea levels, more frequent and more severe storms, increased flooding and drought, and other dramatic changes in weather patterns. Economists estimate that the eventual damage is likely to be substantial. Estimates by the IPCC indicate that a doubling of greenhouse gas (GHG) concentrations would reduce world GDP by 1.5 to 3.5 percent by the end of this century, with economic activity declining by 1 percent in developed countries but significantly more in developing countries, whose economies depend heavily on agriculture.[5] Dell, Jones and Olken (2008) finds significant negative effects of higher temperatures concentrated in poorer countries—a 1 degree Celsius rise in temperature in a given year reduces economic growth in that year by about 1.1 percentage points. The authors find these effects result not only from lower output levels due to agricultural impacts but also from reduced growth due to lower industrial output and aggregate investment and more political instability. Beyond these somewhat predictable developments lies the potential for low-probability but massively catastrophic outcomes.[6]

Energy Security

The problem of energy security is harder to define and often misunderstood. Ever since President Nixon first called for "energy independence," policymakers of all ideological stripes have trumpeted this goal.[7] Between 2001 and 2006, the number of media references to "energy independence" jumped eightfold.[8]

Yet the current obsession with "energy independence," which emphasizes dependence on foreign oil, distracts from the underlying problems of macroeconomic security, household budget security, and national security. The most significant problem with oil, for example, is not that the United States imports it. Because oil is a global commodity that can be shipped at a cost that is low relative to its value, the price of oil is essentially determined by the world market regardless of where it is produced. Thus, even countries that are able to meet all their oil needs through domestic production suffer the economic impacts of price shocks due to global factors. Indeed, the United Kingdom is a net exporter of oil,[9] yet it has seen gas prices rise just as much as has the United States in the past year.[10] As Pulitzer Prize–winning energy expert Dan Yergin put it, "There is only one oil market. . . . Secession is not an option."[11] Rather, the key problem is that U.S. dependence on oil (both imported and domestically produced) exposes the macroeconomy to harmful price shocks and puts long-term strain on household budgets unable to cope with rising costs and price volatility. Moreover, it empowers some of the nation's strategic adversaries and constrains its options for dealing with antagonistic producers. All these risks are exacerbated by the fact that much of the world's oil is concentrated in a few volatile regions (see figure 9-1). As we will show, to increase macroeconomic security and reduce U.S. susceptibility to oil price shocks, the goal must be to reduce consumption of oil and eventually natural gas, regardless of their source.

Though less of an energy security concern for the United States than oil is at present, natural gas is likely to become a growing concern going forward.[12] As utilities have largely been precluded from building new coal-fired power plants or nuclear power plants, 80 percent of the increase in U.S. electricity production since 2000 has come from natural gas–fired plants.[13] Currently, there is no single global price for natural gas, as there is with oil, because gas is distributed regionally. As natural gas exports become increasingly transportable in liquefied form,[14] however, prices will converge toward a single global price. Thus, as the share of U.S. natural gas that is imported in liquefied form rises—roughly threefold by 2030[15]—

F I G U R E 9 - 1 . **World's Proven Oil Reserves, by Region**[a]
Billions of barrels

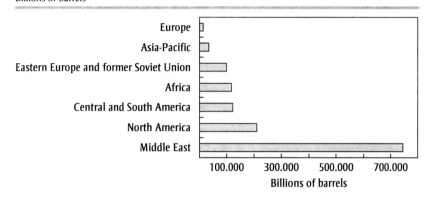

Source: PennWell Corporation, *Oil and Gas Journal*, December 24, 2007.
a. The oil reserve estimate for North America is significantly higher than other estimates. It reflects a much higher oil reserve estimate for Canada that includes an estimated 174 billion barrels of oil sands reserves, for a total of 179 billion barrels. By contrast, the BP Statistical Review of World Energy 2005 estimate of Canadian oil reserves, which is 16.5 billion barrels, includes only the official Canadian estimate of oil sands "under active development." World Oil's 2005 estimate of Canadian oil reserves, which is 12 billion barrels, includes only "reserves that are recoverable with current technology and under present economic conditions," which total only 7.6 billion barrels of oil sands and bitumen.

natural gas dependence will begin to give rise to some of the same concerns that exist with oil: that demand spikes or supply disruptions anywhere will affect prices everywhere.[16] Unlike with oil, however, recent shale discoveries and technology breakthroughs suggest that an abundant supply of natural gas exists in the United States, as noted below. Concerns about natural gas dependence are not unfounded. On the demand side, worldwide natural gas consumption is projected to rise by more than 50 percent by 2030.[17] On the supply side, natural gas supplies are concentrated in a few volatile regions, primarily Russia, Iran, Qatar, Saudi Arabia, and the United Arab Emirates (see figure 9-2). In Europe, for example, the question of gas security already looms large, as Russia accounts for one-quarter of European supplies. The recent conflict between Russia and Georgia, through which a natural gas pipeline has been proposed, also raises questions about the security issues involved with natural gas.

Recent development of unconventional gas sources, including shale gas, has increased estimated reserves of natural gas in the United States significantly. Total proved reserves rose 13 percent in 2007 as a result of development of resources like the Marcellus Shale, which may yield up to 50 trillion cubic feet of natural gas—a large amount relative to annual U.S.

FIGURE 9-2. World's Proven Natural Gas Reserves, by Region
Trillions of cubic feet

Trillions of cubic feet

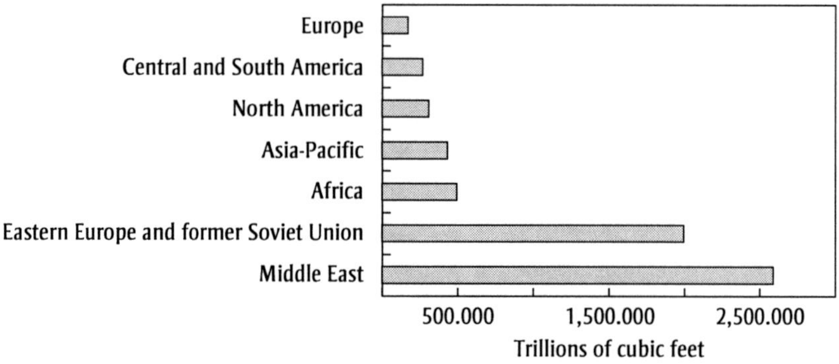

Source: PennWell Corporation, *Oil and Gas Journal*, December 24, 2007.

production of 19 trillion cubic feet per year.[18] These discoveries suggest that the United States may be able to sustain current production levels for several years to come.

Oil and natural gas are not the only energy-related concerns, of course. Nuclear energy raises a host of national security concerns related to proliferation. Market integration of coal has increased in the last few decades, meaning that supply disruptions in one region could affect coal prices in other regions.[19] Other energy issues, such as the reliability and vulnerability of U.S. energy infrastructure (for example, the transmission grid and power plants) also raise security concerns.

But we focus on oil and natural gas because these are the main energy security concerns under our interpretation of the term. Energy security is often defined broadly, with reference to a variety of goals that may align or, in fact, may be in conflict. For example, the term is frequently defined as "access to secure, adequate, reliable, and affordable energy supplies." In our view, however, energy security is more precisely defined around two primary components: a macroeconomic component that relates largely to upward price shocks and a national security component that relates largely to the geographic locations with the largest fuel reserves. Oil and natural gas are the fuels most concentrated in unstable regions and are thus most susceptible to macroeconomic and national security problems. The coal market, in contrast, is far less integrated and less concentrated in unstable regions: less than 20 percent of proven coal reserves are found in

FIGURE 9-3. Dollar Value of Crude Oil Consumption as a Percentage of Nominal U.S. GDP, 1970–2007 Annually and 2008 Q1 and Q3

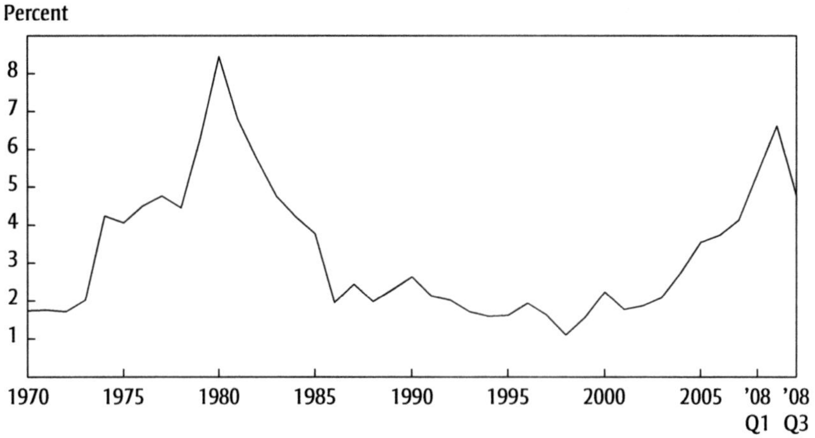

Percent

Source: EIA (2008i), Federal Reserve Bank of St. Louis (2008), and BEA (n.d.), table 1.1.5

the Middle East and Russia, compared with more than 70 percent of reserves of both oil and natural gas.[20] We elaborate on each component of the energy security problem below.

Macroeconomic Security Concerns   At a macroeconomic level, energy price shocks can have harmful impacts, disrupting firms' usual methods of production and reducing households' purchasing power, which can trigger drops in consumer confidence and concomitant reductions in economic activity.[21] Higher energy prices can also feed into higher prices of other goods and thereby induce contractionary monetary policy.[22] For example, nine of the ten U.S. recessions since World War II were preceded by upward spikes in oil prices.[23] Economist James Hamilton finds that there would not have been a recession in the United States from the fourth quarter of 2007 to the third quarter of 2008 without the oil price spike (see figure 9-3).[24] Furthermore, increased oil imports have contributed to the U.S. current account deficit, which is not sustainable in the long run and can slow economic growth.[25] To be sure, lower energy intensity, improved management of monetary policy, and greater flexibility of the economy are all among the factors that have decreased the economy's vulnerability to oil shocks,[26] but the vulnerability is still quite real.

Exacerbating the macroeconomic costs are three market failures that lead individuals to fail to consider the social impact of their private consumption decisions.[27] First, market imperfections, such as wage and price rigidities and producers' inability to adjust energy inputs rapidly, decrease the economy's capacity to adjust to price shocks, leading to excessive underutilization of available resources and increasing aggregate economic costs during price spikes. While mechanisms to help businesses and households adjust more easily (such as oil futures markets and energy conservation measures) have proliferated in recent years, consumers and producers will take protective actions against only the risks that they expect to bear directly, failing to consider the broader impact of their decisions. In addition, such hedging mechanisms are clearly not available or accessible to every household. Second, imperfect competition in the world oil market due to OPEC's market power may inflate the oil price above its competitive level. Since the United States consumes one-quarter of the world's oil production, a reduction in U.S. demand could decrease the world oil price and therefore decrease excess wealth transfers to foreign producers. Third, some studies find that reliance on oil leads the United States to spend resources to secure supplies in volatile regions.[28] That extra expenditure increases macroeconomic fragility by increasing the deficit or diverting resources from other worthy investments. Such defense costs, as well as the macroeconomic adjustment costs and the market power costs, are costly to society but remain hidden from individual consumers when they make daily decisions about energy consumption.

Of course, no one should be confident in predicting the direction of oil prices, which is why estimates vary so widely.[29] Prices may decline or at least grow more slowly as projects to extract oil from the deep sea or oil shale become more viable at today's oil price levels. The very low oil prices of the 1980s and 1990s also led to decreased investment and a concomitant shortage of people, skills, and equipment that led the costs of developing oil and gas fields to double from 2005 to 2008,[30] but those barriers to production may be short-lived as high prices lead to increased investment. On the other hand, demand is expected to continue growing briskly over the long run in developing countries, putting upward pressure on global prices. In China, for example, oil and natural gas demand has risen by 9 percent and 15 percent annually, respectively, since 2001. And oil demand is expected to rise in China by 5.7 percent in 2009.[31] China's vehicle fleet alone is slated to grow from 37 million today to more than 370 million in 2030.[32]

National Security Concerns   The macroeconomic security risks posed by the level and volatility of oil prices not only are harmful in their own right but also create serious national security threats to the United States. There are at least four distinct but related national security problems that arise from U.S. dependence on oil, each of which is more effectively addressed by reducing consumption than by reducing imports.

First, U.S. dependence on oil limits foreign policy options because oil-producing states can exert power over the U.S. economy. With roughly 80 percent of the world's oil reserves controlled by national oil companies,[33] the suppliers of the lifeblood of the U.S. economy are some of the very governments that the United States must negotiate with on military threats and human rights. Because the U.S. economy is susceptible to oil shocks regardless of where oil is purchased, the leverage of oil-supported authoritarian governments is present even if the country imports no oil from them. Indeed, it is telling that Iran continues to use its leverage over oil markets in international negotiations even though the United States has not imported a drop of Iranian oil in twenty-five years.[34]

Second, there are national security risks to the U.S. military involvement in the Persian Gulf, which has been driven for more than fifty years in part by U.S. dependence on oil. The U.S. presence in the region not only diverts scarce resources and exposes U.S. forces to attack but also fuels political resentment against the United States, which terrorist organizations have exploited as a recruiting tool.[35] Of course, even if the United States decreases its oil consumption, it may maintain its presence in oil-rich areas because of other strategic interests or to defend the interests of allies that continue to consume oil.

Third, energy-producing foreign governments reap the financial benefits of high prices supported by U.S. demand, raising concerns that they may use that wealth to pursue political objectives hostile to those of the United States. For one, some argue that resource wealth helps prop up authoritarian governments in countries such as Iran, Venezuela, and Russia that may pursue policies inimical to U.S. national security.[36] In addition, of the estimated $3 trillion in non-pension sovereign wealth fund (SWF) assets, for example, roughly 72 percent comes from natural resources.[37] In effect, U.S. spending on oil and gas is being recycled as SWF investments. However, although those investments may pose risks if not subject to adequate safeguards,[38] it is important to recognize that large current account surpluses in China and OPEC countries have enabled the United States to sustain healthy domestic investment and consumption despite a falling national saving rate. Indeed, for the most part, foreign direct investment and foreign purchases of U.S. government debt are vital and beneficial to the U.S. economy.

Fourth, as Thomas Friedman has pointed out, high oil prices are correlated with worrisome domestic impacts in what he calls "petrolist" countries, eroding "free speech, free press, free and fair elections, an independent judiciary, the rule of law, and independent parties."[39] Resource-rich countries are vulnerable to the "resource curse," according to which countries with more resource wealth tend to have lower rates of growth and greater corruption and government mismanagement. Natural resource wealth can lead to conflict in a country, make governments that do not rely on tax revenues less accountable, and make a country's manufacturing sector less competitive by raising the real exchange rate (so-called "Dutch disease").[40]

The latter two concerns are the result of high world prices, which could in theory be addressed by increasing domestic supply rather than (or in addition to) reducing consumption. But reducing consumption is preferable for at least three reasons. First, reducing consumption would have significant ancillary climate benefits, while increasing domestic supply would exacerbate the climate problem. Second, in a market in which demand is more elastic in the long run than supply, a reduction in demand will have a greater negative impact on world prices than an increase in supply of equal magnitude. Third, because the future economic strength of the United States will likely depend on its ability to be a leader in technology innovation, there is greater advantage in developing alternative energy sources and products that consume less energy than in increasing the supply of traditional energy sources. Innovative ways to reduce consumption are more sustainable and have the potential to spread quickly, which is part of the reason that energy demand is elastic in the medium-to-long run, while developing new energy sources may well need a long period of development and cannot spread as quickly as can the diffusion of knowledge and information.

Energy Affordability

The issue of energy security has risen rapidly to the top of the political agenda largely in response to record-high gasoline prices through the summer of 2008, which have imposed pain on many U.S. households. At the consumer level, rising oil prices are putting an increasing strain on household budgets. Total consumer expenditures on oil and gasoline rose 98 per-

FIGURE 9-4. Spending on Gasoline and Motor Oil, by Household Income

Percent of after-tax income

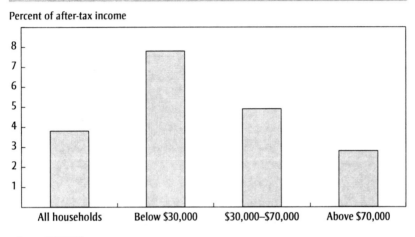

Source: BLS (2006).

cent between 2003 and 2008.[41] Moreover, the impact of higher gas prices is especially acute for low-income households and those in rural areas, who drive more on average and lack alternative means of transportation. In 2006, U.S. households spent on average 4 percent of their income on gasoline, but the number is nearly 8 percent for households with incomes under $30,000 and nearly 5 percent for those with incomes between $30,000 and $70,000 (figure 9-4).[42] In addition, spending on gasoline as a percent of household income reached the double digits in many areas of the South, Southwest, and Great Plains and even exceeded 13 percent in many poor rural counties where there are few alternatives to driving long distances.[43]

Moreover, the problem is not only high oil prices but also unexpected fluctuations in oil prices. Predictability is crucial to long-term, sustainable affordability. Even if prices are relatively low, consumers are harmed if they are volatile because households are unable to plan or budget adequately. For example, many who purchased less fuel-efficient vehicles when gas was cheap found those cars to be a serious financial burden. Long-term household budget security would be enhanced if people had more predictability about the long-term operating costs related to their consumption decisions.

In addition to higher oil prices, Americans are increasingly concerned about higher energy prices because of the widespread expectation that Congress will act in the near future to address climate change by putting a price on carbon, most likely through a cap-and-trade system. According to figures estimated in Metcalf (2007), a $15 per ton price on carbon dioxide (CO_2) would increase electricity prices by nearly 20 percent, gasoline prices by 3 percent, and natural gas prices by 3 percent over 2008 prices.

Managing the Interactions between Climate and Energy Policy

Not only is there tension between the goal of energy affordability and the goals of reducing climate change and oil consumption, which may require higher rather than lower prices, but the issues of energy security and climate change themselves are often misleadingly conflated. Many assume that addressing U.S. energy security needs will also promote progress on climate change, and vice versa. To be sure, in many ways, energy security and climate change do overlap. For example, petroleum accounts for 44 percent of U.S. carbon emissions, so reducing oil use would have significant climate benefits.[44] But in at least four less appreciated respects, climate change and energy security are distinct and, at times, even conflicting goals.

First, there is a temporal difference between the problems. Climate change is a long-term problem since carbon stays in the atmosphere for so long, while energy security is a much more immediate concern. Successful climate change policies can thus phase in emissions reductions over time and include cost-containment mechanisms since the marginal benefit of reducing emissions in any given year is low while the cost of meeting any given year's cap may be high.[45] In contrast, the risks associated with energy security already threaten the U.S. economy and limit its foreign policy options, thereby requiring more immediate reductions in oil use. The effect of the oil price shock of 1973, for example, was an immediate decrease in the real GDP of the U.S. economy for the subsequent two years and an increase in the amount that households paid for gasoline. Similarly, the economy would not have been in recession from the last quarter of 2007 through the third quarter of 2008 had it not been for the oil price shock.[46] Despite the immediacy of the problem, the solution to energy security, as we propose later, may need to be gradual to avoid sudden changes to the tax system or to household budgets.

Second, there is a geographic difference. The United States can take unilateral measures to promote its energy security by decreasing its oil consumption and thus the oil intensity of its economy and its susceptibility to oil price shocks. But climate change is an inherently global phenomenon, as emissions from any part of the world make an equal contribution to the problem. Any successful climate policy is therefore impossible without international cooperation, primarily from major GHG-emitting nations like China. To put in perspective the importance of global action, consider that if Chinese CO_2 emissions continue to grow at the rate that they did from 2002 to 2007, then by 2036 China *alone* will consume the entire CO_2 emissions budget that would be allowed to reach a 450 parts-per-million global CO_2 stabilization goal.[47]

Third, whereas reducing oil and gas consumption would address their relevant energy security concerns, reducing carbon emissions (that is, mitigation) will not be enough on its own to tackle the problem of climate change. Current and past emissions will continue to warm the planet over the next century even if all future emissions are avoided. Even as governments attempt to reduce the "flow" of new emissions into the atmosphere, the planet is already experiencing climate change from the "stock" of carbon emissions in the atmosphere. A comprehensive approach to addressing climate change therefore must include not only *mitigation* of future emissions but also *adaptation* to the consequences of past emissions.[48]

Finally, while reducing oil and natural gas consumption is the primary way to alleviate energy security concerns as we have defined them (that is, energy problems with both macroeconomic and national security dimensions), reducing coal consumption is the primary way to alleviate climate-change concerns. One reason that coal consumption is more likely to fall as a result of climate policies is that there are more low-cost alternatives to coal used in the electricity sector than there are to oil used in the transportation and manufacturing sectors. That is compounded by the fact that compared with coal, oil has relatively little carbon per unit of energy. In addition, the carbon component of coal power is a higher proportion of the final price than the carbon component of oil used in gasoline, where about $1 per gallon goes to taxes, marketing, refining, and distribution.[49] As a result, putting a price on carbon emissions—which is the core of a cost-effective climate change strategy—would result in a much larger increase in the price of coal than in the price of gasoline. For example, a CO_2 tax of $25 per ton would increase the price of oil, starting at a base price of $100 per barrel, by about 10 percent (a hike of roughly $0.24 per gallon at the pump), but it would roughly double the price of coal.[50]

Carbon policies may therefore do relatively little to curb oil use, especially given the cost of alternatives and of reducing consumption outright, and certainly they would not induce oil demand reductions on their own that would be large enough to seriously improve energy security. Indeed, by sharply raising the price of coal relative to that of natural gas, a cap-and-trade system or carbon tax may actually *increase* energy security risks by causing a substitution of natural gas for coal.[51]

Because climate change and energy security concern different fuels to different degrees, efforts to make progress on one may come at the expense of the other. Policymakers need to minimize such conflicts. Coal use is one example of the tradeoff, since coal has fewer energy security risks than oil or gas but greater climate risks. If carbon capture and storage (CCS) proves feasible, then the United States may be able to continue consuming coal at current levels with minimal impact on the climate. But the viability and safety of CCS technology have yet to be proven on a large scale.[52] Perhaps the clearest example of the trade-off between energy security and climate change is coal-to-liquid technology, in which coal is transformed into a diesel fuel that can be used in place of conventional oil. Although this technology would promote energy security by reducing oil consumption, it would exacerbate climate change by emitting more than twice the amount of GHGs as conventional oil production, in addition to requiring intense coal mining and large amounts of water.[53] Similarly, subsidies for ethanol may reduce oil consumption, but recent evidence suggests that corn-based ethanol may actually lead to more GHG emissions when land use changes are taken into account.[54]

The recent run up in oil prices also reveals the ways in which climate change and energy security can come into conflict due to unintended consequences. From a climate change perspective, high oil prices were partly a welcome development because of the incentives created to reduce oil consumption. However, high oil prices have also made it profitable to extract hard-to-recover and dirtier fossil fuels like oil shale and to use coal-to-liquids technology.[55] Similarly, since natural gas prices usually track crude oil prices,[56] higher natural gas prices have led some European utilities to calculate that burning coal, even with the higher carbon charge in the EU's cap-and-trade system, is cheaper than burning cleaner natural gas.[57] On the flip side, a high enough carbon price may exacerbate energy security concerns if it results in a switch from coal to natural gas, increasingly in the form of liquified natural gas (LNG) from unstable regions.

A Cost-Effective Response to Climate, Energy Security, and Affordability Challenges

The optimal policy responses to climate change, energy security, and energy affordability leverage the ways in which these challenges overlap while taking into account the ways in which they may conflict to avoid making progress on one at the expense of the others. The core of a cost-effective response to the first two issues is the use of a market mechanism (a cap-and-trade system or a carbon tax) to reduce demand and encourage fuel substitution. As discussed below, these mechanisms are generally preferable to command-and-control regulations, although additional policies such as efficiency standards and research subsidies may still be needed to respond to certain market failures to lower costs and increase affordability. We discuss the advantages of market mechanisms aimed at reducing carbon emissions and oil consumption in turn. While the market mechanisms promote price stability and long-run affordability, they also lead to higher energy prices, at least in the short term. Therefore, we also recommend using the revenue raised by such mechanisms to offset their adverse distributional impacts. Finally, we propose cost-effective policies such as subsidies for basic research and development, better consumer information, and other reforms aimed at increasing energy affordability and reducing the cost of mitigating the negative externalities from carbon and oil consumption.

Climate Change

In the case of climate change, there is a growing consensus that a market mechanism such as a cap-and-trade system or a carbon tax should be at the heart of any effort to reduce GHG emissions.[58] The major advantages of market mechanisms are innovation, flexibility, and cost effectiveness. The basic intuition underlying the claims for their benefits is that firms and individuals know better than the government where the most cost-effective reductions are likely to occur. Firms searching for methods to reduce emissions in order to avoid paying the tax or using permits will have numerous ways to reduce emissions, from changing production processes to shifting the sources of energy or raw materials. Firms that figure out the most cost-effective ways to accomplish that task would succeed, and, in a competitive economy, other firms would either have to copy their best practices or cease to exist. Given the proper incentives, the decentralized decisions of profit-maximizing firms would lead to substantial innovation and ingenuity in curbing carbon emissions—well beyond anything that regulators could envision.

Firms would pass on most of their increased costs to consumers, who would respond to the higher prices by adjusting their behavior.[59] The mix of solutions would vary from person to person, based on each individual's tastes and personal circumstances. In response to a carbon price, for example, some individuals may wish to take public transportation rather than drive, while others may place greater value on driving and thus prefer to reduce emissions by using more energy-efficient light bulbs or through other means. A carbon price signal allows individuals to make the decisions that maximize their welfare. The flexibility of price mechanisms would allow both consumers and firms to make the most cost-effective choices in response to price signals.

Harvard economist Rob Stavins gathered estimates of the macroeconomic impact of market mechanisms and concluded that an efficient cap-and-trade system that stabilized carbon emissions at their 2008 level by 2050 would cost less than 0.5 percent of GDP in each year of the program (and just 1.2 percent for a more aggressive climate policy).[60] The cost of a cap-and-trade system can also be significantly reduced, at little or no harm to the environment, through cost-containment mechanisms such as a safety valve and banking and borrowing of permits across years, which can minimize short-term fluctuations in permit prices.[61] Compared with other policies, use of such a market mechanism can significantly reduce the cost of achieving carbon reductions. Comparing the cost of achieving the same 5 percent emission reduction in the electricity sector under various policies, Fischer and Newell (2007) estimates that it would be twice as costly to do so with a renewable portfolio standard as with an emissions price and twelve times as costly to do so through subsidies for research on renewable sources of energy.

Energy Security

Just as pricing carbon is the most efficient way to mitigate climate change, the most cost-effective approach to achieving energy security would be to put a price on oil that forces consumers to internalize the social harm caused by their oil use, over and above the impact on oil prices of putting a price on carbon. To the extent that natural gas consumption begins to impose macroeconomic and national security costs on society, pricing natural gas to reflect those social costs would also be the most cost-effective policy response. We focus here on pricing oil, but the same logic would apply to a system that was intended to increase energy security by discouraging natural gas consumption. We focus on oil in part because the social costs associated with oil consumption are more apparent, while the energy security costs of natural gas consumption appear ambiguous, at least for now. While an oil price is the most efficient way of reducing macroeconomic and national security risks from oil consumption, it will have important distributional consequences. Later in the chapter we outline a proposal to use the revenue raised to offset the price impact in a progressive manner.

The goal of pricing oil is to reduce the macroeconomic and national security social costs of oil dependence by inducing a reduction in U.S. oil consumption. As with climate change, the size of the oil price should be linked to the social costs of oil consumption. Estimates vary on how large an optimal tax on oil should be. A range of studies estimate the economic externalities of oil to be about $20 per barrel (about $0.50 per gallon of gasoline).[62] Although widely recognized, the external national security costs of oil dependence[63] are inherently difficult to quantify in dollar terms, and we do not know of any studies that endeavor to do so. More research certainly is needed to try to quantify the cost. If we assume that the national security costs of oil dependence are comparable in magnitude with the economic costs, the combined external cost of oil consumption would be roughly $40 per barrel, or $1 per gallon of gasoline, which is the amount of the gasoline tax advocated, for example, by Greg Mankiw, former chairman of President Bush's Council of Economic Advisors (CEA).[64] A tax on the order of $1 per gallon would still be well below the gas taxes that exist in many other OECD nations, which are roughly four to five times that amount (see figure 9-5).[65] Though many have proposed higher taxes on gasoline,[66] the external costs associated with oil dependence justify a tax on all oil, not just the 69 percent of oil consumed in transportation fuels.[67]

The purpose of an oil tax, environmental gains aside, is to make consumers bear the full social cost of their decisions, thereby reducing oil consumption and the oil intensity of the U.S. economy and thus the nation's vulnerability to future price shocks. Paradoxically, however, as a political matter, periods of oil price spikes are the worst time to propose oil taxes. Higher oil prices are widely seen as the problem, not the solution. For that reason, we propose an oil tax that phases in very gradually but does so more quickly if oil prices decline. Such a variable tax would function as a hybrid between a tax and a price floor, providing both greater price sta-

F I G U R E 9 - 5 . Gasoline Prices and Taxes, by Country

Total price at pump, July 2008 (dollar/gallon)

Source: IEA (2008a).

bility and an incentive for consumers to reduce consumption. Specifically, we propose a tax on oil with a magnitude that varies depending on the price of a barrel of oil. For example, a tax on oil might start low at $2 per barrel and phase up gradually to $40 over twenty years in order to give consumers and producers some lead time to develop substitutes for their current energy consumption patterns. But then it might rise more steeply in times of oil price declines and more slowly in times of oil price increases. When oil prices fall by a dollar per barrel, for example, the tax might rise by 50 cents more than it otherwise would; conversely, when oil prices rise by a dollar, the tax might rise 50 cents less than it otherwise would. The eventual $40 per barrel tax[68] would add about $1 to a gallon of gasoline, in increments of about $0.05 per year.

Tables 9-1 and 9-2 show how such a variable oil tax would be applied in different scenarios. As seen in table 9-1, if the price of oil were to hold steady, the tax would rise to $10 per barrel after five years. If the price of oil were to rise from $140 to $180 over that period, the tax would be $0

TABLE 9-1. Retail Price of Oil with Variable Oil Tax and Rapid Changes in Price[a]

Year	Scenario 1: Price of oil steady at $140/barrel			Scenario 2: Price of oil rises to $180/barrel over five years			Scenario 3: Price of oil falls to $100/barrel over five years		
	Price of oil ($/barrel)	Tax ($/barrel)	Retail price ($/barrel)	Price of oil ($/barrel)	Tax ($/barrel)	Retail price ($/barrel)	Price of oil ($/barrel)	Tax ($/barrel)	Retail price ($/barrel)
0	140	0	140	140	0	140	140	0	140
1	140	2	142	148	0	148	132	6	138
2	140	4	144	156	0	156	124	12	136
3	140	6	146	164	0	164	116	18	134
4	140	8	148	172	0	172	108	24	132
5	140	10	150	180	0	180	100	30	130

Source: Authors' calculations.

a. The proposed tax increases by $2 per barrel per year until a maximum of $40 per barrel is reached. If the price per barrel falls by $1, the tax increases by $0.50 more than otherwise; if the price per barrel rises by $1, the tax increases by $0.50 less than otherwise. The average increase of $2 per barrel per year equals about $0.05 per gallon of gasoline per year. The eventual $40 per barrel maximum tax equals about $1 per gallon of gasoline.

TABLE 9-2. Retail Price of Oil with Variable Oil Tax and Gradual Changes in Price[a]

Year	Scenario 1: Price of oil steady at $140/barrel			Scenario 2: Price of oil rises to $180/barrel over five years			Scenario 3: Price of oil falls to $100/barrel over five years		
	Price of oil ($/barrel)	Tax ($/barrel)	Retail price ($/barrel)	Price of oil ($/barrel)	Tax ($/barrel)	Retail price ($/barrel)	Price of oil ($/barrel)	Tax ($/barrel)	Retail price ($/barrel)
0	140	0	140	140	0	140	140	0	140
1	140	2	142	142	1	143	138	3	141
2	140	4	144	144	2	146	136	6	142
3	140	6	146	146	3	149	134	9	143
4	140	8	148	148	4	152	132	12	144
5	140	10	150	150	5	155	130	15	145
6	140	12	152	152	6	158	128	18	146
7	140	14	154	154	7	161	126	21	147
8	140	16	156	156	8	164	124	24	148
9	140	18	158	158	9	167	122	27	149
10	140	20	160	160	10	170	120	30	150
11	140	22	162	162	11	173	118	33	151
12	140	24	164	164	12	176	116	36	152
13	140	26	166	166	13	179	114	39	153
14	140	28	168	168	14	182	112	40	152
15	140	30	170	170	15	185	110	40	150
16	140	32	172	172	16	188	108	40	148
17	140	34	174	174	17	191	106	40	146
18	140	36	176	176	18	194	104	40	144
19	140	38	178	178	19	197	102	40	142
20	140	40	180	180	20	200	100	40	140

Source: Authors' calculations.

a. The proposed tax increases by $2 per barrel per year until a maximum of $40 per barrel. If the price per barrel falls by $1, the tax increases by $0.50 more than otherwise; if the price per barrel rises by $1, the tax increases by $0.50 less than otherwise. The average increase of $2 per barrel per year equals about $0.05 per gallon of gasoline per year. The eventual $40 per barrel maximum tax equals about $1 per gallon of gasoline.

(because the proposed tax increase of $2 per year is completely offset in any year in which the price of oil rises by $4 or more since the tax increase is reduced by $0.50 for every $1 rise in the price of oil). By contrast, if the price of oil were to decline from $140 to $100 over the same period, the tax would reach $30 per barrel, and thus the after-tax price would decline by only $10, from $140 to $130. If the price of oil were to rise much more gradually, say over twenty years rather than five, the tax would be $20 in year 20, bringing the after-tax price to $200. By contrast, if the price were to decline from $140 to $100, the tax would be $40, bringing the after-tax price back to $140.

Our proposed formula is intended only to be illustrative; the actual formula could be different in myriad ways. For example, the tax could rise faster, say in $4 per barrel increments rather than in $2 per barrel increments. Or the cushion for high oil prices could be more generous, with the tax falling $0.75 instead of $0.50 for every $1 rise in oil prices. The important point is that an oil tax that rises very slowly, if at all, when the price of oil goes up but rises much more sharply when the price of oil declines can be a pragmatic way to provide price stability and reduce oil consumption and the oil intensity of the economy while minimizing the economic pain an oil tax would impose on U.S. households.

To be sure, as a purely theoretical matter, the optimal amount of the tax should be determined by the external costs of oil dependence, not by the market price for oil in the way that our proposed variable oil tax is.[69] Further, as mentioned, energy security is an immediate problem that in theory calls for a short-term solution. But the sort of gradual hybrid solution described is more appropriate for balancing the goal of energy security with the goal of stability in the tax system and in household budgets. First, a gradual tax gives consumers time to adjust their lifestyles in response to a sustained price signal. Second, given the finding that the pain that people perceive from forgoing a dollar's gain is significantly less than the pain from losing a dollar,[70] an oil tax that prevents prices from declining more than it increases them may be a more acceptable approach. It is important to note that this research implies that the approach may also be not just politically but also economically more efficient since it achieves a given tax level by reducing the gains from lower prices rather than increasing the loss from higher taxes.

That a tax would be necessary at all if the price of oil goes up may be counterintuitive because, as recent experience has demonstrated, a large increase in the market price itself would reduce oil consumption. Yet although it is true that higher market prices will lower consumption, the purpose of an oil tax is not to achieve a certain level of consumption. Rather it is to get consumers to make socially appropriate consumption decisions by sending them price signals commensurate with the costs that their consumption is imposing on society in general. When the market price of oil rises, the macroeconomic and national security external costs of oil consumption do not disappear. In fact, there is evidence that the economic costs might be even higher when oil prices rise because spending on oil as a fraction of GDP rises and increases the potential cost of price shocks.[71]

Recent evidence from higher gas prices indicates that an oil tax such as the one that we propose would be effective in reducing oil consumption. As a result of recent gas price increases, consumers are driving less, purchasing more fuel-efficient vehicles, and using more public transit. The quantity of gasoline demanded fell 2.2 percent in July 2008 from the level a year earlier, resulting in the lowest July level in four years.[72] Vehicle miles traveled fell by 4.7 percent in June 2008 from the June 2007 figure, the biggest monthly decline from a year earlier since the data began to be recorded in 1983.[73] Transit ridership has surged to its highest level in fifty years, with Americans taking 10.3 billion public transit trips in 2007.[74] According to Cambridge Energy Research Associates, "U.S. gasoline demand will likely decline in 2008 for the first time in more than 17 years."[75] The impact is likely to be even greater in the long term since economic evidence confirms that demand for driving is more sensitive to gas price increases in the long run than in the short run. For example, a 10 percent increase in gas prices would reduce gasoline consumption by just 0.6 percent in the short run but by 4 percent in the long run.[76] By those estimates, an eventual $1 increase in the gasoline tax would reduce consumption by 13 percent in the long term (starting from a base retail price of $3 per gallon). In addition, the long-term response to a tax on oil will be even more significant than the response to high gas prices because a tax is a predictable and sustained long-term price signal. People determining whether it makes good financial sense to move closer to work or buy a more fuel-efficient vehicle, for example, will have more certainty about long-term fuel prices. On the supply side, businesses will also be able to make long-term investments in alternative fuels without worrying that demand will be undercut by a precipitous drop in oil prices.

Finally, there is some debate about whether oil substitutes such as ethanol also should be taxed due to energy security externalities. The logic is that since they are close substitutes, the prices of oil and ethanol tend to move together. An oil price shock will make ethanol relatively more affordable, driving up demand for ethanol and eventually its price. In such a scenario, the U.S. macroeconomy is still susceptible to an oil price shock even if it relies on ethanol instead of oil. To the extent that consumers do not take social costs into account when making individual consumption decisions, the optimal policy response would be a tax to force consumers to internalize those cost. But there are three reasons, from an economic and national security perspective, why substitutes such as ethanol should be taxed at a lower rate than oil. First, ethanol consumption does not have an external national security cost. Second, only part of the macro-economic security externality is relevant; there is no OPEC-like market power in the ethanol market. Finally, even the price volatility component is only as strong as the correlation between ethanol and oil prices, and that correlation will tend to decrease as the market for ethanol develops. In the long run, if a competitive market develops, the price will be driven to the marginal cost of production and the impact of oil price shocks on ethanol prices should dissipate.

Energy Affordability

While a price signal is the most cost-effective way to address climate change and energy security, many consumers can be hurt significantly by higher energy prices. Low-income households spend roughly 14 percent of their income on energy bills, while the national average is only 3.5 percent.[77] In response, four measures can help improve energy affordability: using revenue from a market mechanism to offset the distributional impacts of such policy; mitigating exposure to harmful price volatility; removing market distortions that drive up the pretax cost of energy; and implementing limited and well-targeted R&D subsidies and regulations to address information barriers and other market failures while avoiding less efficient command-and-control approaches.

Offset the Distributional Impact of Market Mechanisms   The cost to consumers of using pricing mechanisms can be offset by distributing the revenue that they generate. For example, to offset the distributional impact of a carbon price, revenues raised from a carbon tax or from auctioning cap-and-trade allowances could be redistributed to individuals, thus maintaining the incentive to reduce carbon use while offsetting increased energy costs. Economist Gilbert Metcalf estimates that the revenue raised from a carbon tax of $15 per ton of CO_2 could be redistributed to every

FIGURE 9-6. Distributional Impact of a Carbon Tax of $15 per ton of CO$_2$ with a Lump-Sum Rebate to All Households

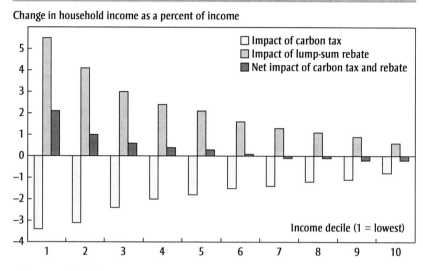

Change in household income as a percent of income

Source: Metcalf (2007).

individual in an equal lump-sum rebate of $274.[78] As figure 9-6 shows, such a lump sum rebate would be a moderately progressive proposal, as very low-income tax payers would end up better off, while most others would end up no better or no worse off.

A similar approach should be taken with revenue from the sort of variable oil tax proposed above. With such a tax, the higher prices that people pay would accrue to the federal government as revenue rather than to oil-producing nations that reap rents from their natural resources. The government can then use that revenue to offset the distributional impacts of an oil tax. Our analysis using data from the 2006 Consumer Expenditure Survey from the Bureau of Labor Statistics shows that a $1-per-gallon tax on gasoline and motor oil would raise $101 billion. If that revenue were returned in an equal lump-sum tax rebate of $850 to each household in the United States, households in the bottom three quintiles would be better off on average while those in the top two quintiles would be marginally worse off (see figure 9-7).[79]

Mitigate Exposure to Price Volatility   Second, policymakers should mitigate susceptibility to energy price volatility rather than strive for short-

F I G U R E   9 - 7 . Distributional Impact of a Gasoline and Motor Fuel Tax of $1 Per Gallon with a Lump-Sum Rebate to All Households

Change in household income as a percent of income

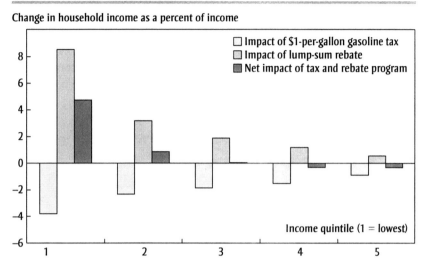

run, short-lived affordability, as unexpected price shocks put significant strains on household budgets. After the unexpected oil price shocks of the late 1970s, for example, average consumer expenditures on gasoline and fuel oil rose to more than 5 percent of household income, up from less than 3.5 percent of income just a decade earlier. A variable oil tax would help achieve this goal by partially curtailing price reductions today that would hinder the behavioral changes and innovations necessary to reduce the oil intensity of the U.S. economy and thus the country's exposure to future oil price shocks. Much like a vaccine prevents a painful disease outbreak in the future by exposing people, in a controlled way, to low levels of the virus today, artificially increasing the price of oil today, in a predictable and gradual manner, can reduce the pain of oil price volatility in the future. Modestly higher prices—particularly prices that are stable and predictable and result from preventing price declines rather than raising prices—can be less painful than unexpected price shocks, which require more wrenching and rapid adjustments. Modestly higher prices today also enhance intergenerational equity. Just as current generations should internalize the cost of their carbon emissions even though the effects of climate change will mostly be felt by future generations, beginning to reduce oil consumption today will make the ineluctable transition away from that finite natural resource less costly and painful for future generations.

In the case of a carbon price, price volatility can be mitigated through the use of cost-containment mechanisms such as a safety valve or reserve allowance. The nature of climate change allows for flexibility in the timing of emissions reductions since the harm of climate change comes from the long-term stock accumulation of CO_2 in the atmosphere; the degree of warming responds very slowly to changes in the annual level of emissions. Economic costs, however, are very sensitive to the emissions level. Meeting a certain quantity cap may turn out to be quite costly in a given year even though the gain from reducing emissions in that particular year rather than a future year is quite small. A cost-containment mechanism can thus significantly reduce energy price volatility for consumers at very little environmental cost.

Lower Energy Prices in Economically Sound Ways   Third, while an oil tax would raise prices, certain economically sound policies can also reduce prices by making the pretax private price equal the free-market price to the greatest degree possible. Keeping the private market price in line with costs means eliminating artificial barriers to increased supply and reduced demand. For example, supply should be increased to the extent that it can be done in an environmentally safe manner and the potential reserves justify any risks.[80] The United States should also eliminate the $0.54-per-gallon tariff on imported ethanol. Since federal and state mandates require ethanol to be blended with gasoline, any policy that constricts the supply of ethanol increases retail gasoline prices. This policy is especially harmful because imported ethanol comes mostly from Brazil, where ethanol is made from sugarcane in a significantly more environmentally friendly manner than the process used to make ethanol from corn in the United States. The short-run impact is not likely to be significant, however, because ethanol imports would take time to ramp up. A recent study from Iowa State University found that eliminating the import tariff would reduce the retail price of gasoline by $0.03 per gallon between September 2008 and September 2009.[81]

On the demand side, the United States should encourage other governments to reduce or eliminate fuel subsidies. Today, roughly a quarter of the world's oil is sold at below-market prices in countries that subsidize fuel costs.[82] These market distortions prop up demand and keep oil prices at artificially high levels.[83] Indeed, on the day in June 2008 that China raised its base price for gasoline by 17 percent, the price of benchmark crude oil fell by 3.5 percent as traders anticipated a concomitant reduction in Chinese demand.[84]

Such efforts should be distinguished from the myriad solutions recently proposed to reduce short-term oil prices—whether to crack down on speculators, curb inventories in the Strategic Petroleum Reserve, or grant a gas-tax holiday. Each of those policies would have a negligible effect on gas prices and in some cases would be harmful. Fundamentally, high oil prices are the result of rapidly growing global demand and supplies that have not kept pace, a situation made worse by the weak dollar and geopolitical fears of supply disruptions.[85]

Use Cost-Effective Market Mechanisms and Well-Targeted Subsidies and Regulations   Finally, energy affordability can be increased by minimizing the cost borne by individuals to achieve U.S. energy and climate goals with cost-effective climate and energy policies and by using targeted R&D and information policies to help consumers save money. Market mechanisms are the single most cost-effective tools available to achieve energy and climate goals, but limited and well-targeted regulations, if pursued along with comprehensive market mechanisms, can help lower the costs of reducing carbon, oil, and natural gas consumption and increase the long-term affordability of energy.

Generally speaking, command-and-control regulations, mandates, or subsidies tend to raise, not lower, the cost of achieving these goals, and consumers ultimately bear the costs in the form of higher prices for goods that meet new standards or higher taxes to fund government subsidization. There are several reasons command-and-control policies are more costly, less effective, and less efficient. First, command-and-control systems generally cover only a fraction of the economic and behavioral choices that affect emissions or oil consumption. For example, renewable portfolio standards affect only one dimension of choice with regard to one source of emissions—electricity generation—and therefore do not necessarily take advantage of the cheapest way to reduce emissions, even within the electricity sector (see figure 9-8). Second, the government has limited knowledge of the best ways to reduce GHG emissions or oil consumption. Choosing the best way among myriad options would require a sophisticated understanding not only of technology and economics but also of individual preferences. Third, the government has a poor track record in picking winners and losers when choosing to subsidize or otherwise create incentives for particular technologies.[86] Not only do decisions about the next set of technological winners require knowledge about

FIGURE 9-8. Economic Impact of Various Approaches
to Energy Security

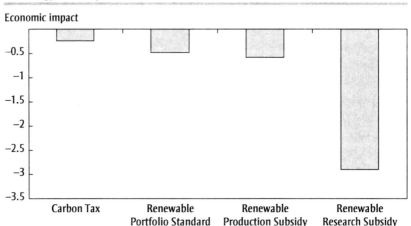

Source: Data from Fisher and Newell (2007).

complex aspects of the economy, as discussed above, but they also are susceptible to political pressures. Consider, for example, the enormous government enthusiasm for corn-based ethanol as a substitute for petroleum. Economist Gib Metcalf, for example, recently estimated that the cost of reducing CO_2 emissions through the tax credit for ethanol exceeded $1,700 per ton of CO_2 avoided in 2006,[87] and subsequent research has found that corn-based ethanol has little environmental benefit and may even be harmful from a climate change perspective.[88]

An example of the inefficiency of command-and-control approaches is CAFE (corporate average fuel economy) standards. Although they have helped reduce oil consumption and GHG emissions, unlike a carbon or oil tax CAFE standards make it cheaper to drive each mile and so encourage people to drive more.[89] Also unlike pricing mechanisms, CAFE does not achieve near-term demand reductions because it is relevant only for the purchase of new vehicles and, by raising the cost of new vehicles, may actually encourage people to drive less fuel-efficient vehicles longer. CAFE may also push consumers from automobiles, which are covered by CAFE, to less fuel-efficient SUVs, which are covered by looser standards, thus demonstrating the risk of unintended consequences with imperfectly set standards. Moreover, higher fuel efficiency standards have diminishing returns. Consider that oil consumption would be reduced more by getting a driver whose cars gets 10 mpg to switch to a vehicle that gets 12 mpg than by getting someone whose car gets 50 mpg to switch to one that gets 100 mpg.[90] What matters is relative differences in *gallons per mile,* not miles per gallon. Switching from 10 mpg to 12 mpg has a greater impact for a given mileage because it generates fuel savings of 1/60 gallons per mile, whereas switching from 50 mpg to 100 mpg only generates fuel saving of 1/100 gallons per mile. That's not to say that in the long term vehicles that use very little gasoline are not going to be a critical part of the solution to the energy crisis, but in the short term it would do more good to get people who drive SUVs to drive sedans than to get those who drive hybrids to drive plug-in electric hybrids. That is why programs that target the most fuel-inefficient cars immediately, like "cash for clunkers," can be so effective.[91]

In certain cases, however, well-targeted and limited subsidies and regulations can help lower costs and increase affordability by addressing specific market failures. Government funding for basic research into new energy technologies, for example, would benefit consumers. Since studies show that innovators capture less than one-quarter of the total value of their innovations, the private sector invests less in R&D than is necessary for the nation to realize the full potential of technological innovations and thus achieve lower-cost alternative energy technologies in the future. Studies show that, of the public investments that are made, energy R&D investments in particular yield substantial economic benefits and lead to significant knowledge creation.[92] A recent Hamilton Project study by Richard Newell argues that doubling energy R&D to roughly $8 billion per year by 2016 is justified given the substantial needs and opportunities for basic energy research.[93]

Certain market failures may also justify limited command-and-control regulations or standards, on top of prices for oil and carbon, to reduce the costs of achieving reductions and increase energy affordability. For example, there may be principal-agent problems that lead to misaligned incentives, such as between those who build buildings and those who inhabit them. Home builders have little incentive to promote energy efficiency because cost savings accrue largely to the building's tenants, who may lack full information about the building's efficiency. Regulations to enhance transparency of information on building efficiency standards may thus be warranted to address such market failures.

In addition, there may be information failures that obscure price signals or leave consumers without the knowledge to respond effectively to price signals. For example, consumers today often lack a clear sense of how energy is priced and the extent of their electricity use or the fuel efficiency of their cars.[94] People may therefore have difficulty taking steps to reduce consumption efficiently in response to higher prices. Relatively inexpensive and efficiency-enhancing mandates to provide information feedback to consumers can leverage insights from behavioral economics to reduce oil use and carbon emissions and save consumers money in the long run. For example, displays that inform consumers in real time about how much electricity they are using have been shown to reduce electricity consumption.[95] Our preliminary calculations show that such devices could save the average U.S. consumer $115 per year on his or her electricity bill and yield a net social benefit of $11.8 billion over ten years. Mandating that vehicles display fuel efficiency data to drivers might similarly lead drivers to improve their fuel economy and thus save them money.[96]

Conclusion

To achieve greater energy security, mitigate climate change, and enhance energy affordability, those from both the left and the right of the political spectrum may be able to find common ground in a grand bargain that uses market mechanisms to reduce demand and develop substitute fuels in the most cost-effective way and helps consumers by expanding energy supplies, redistributing revenues raised through rebates to households, minimizing price volatility, and addressing various market failures. Putting a price on oil is the most cost-effective way to reduce the oil intensity of the U.S. economy, thereby minimizing the nation's vulnerability to the harmful economic impacts of oil supply disruptions and price shocks. A variable oil tax that rises more sharply as prices decline to create a sliding floor on oil prices can achieve this goal with less pain to consumers, who tend to exhibit loss aversion. To address climate change, a carbon tax or cap-and-trade system that auctions permits, thus raising revenue that can be redistributed to consumers, and includes cost-containment mechanisms can best minimize energy costs for consumers and harm to the macroeconomy. Long-term energy affordability can also be enhanced when such measures are coupled with efforts to remove market distortions that artificially inflate the pretax costs of energy and to address specific market failures with well-targeted regulations.

Although the approaches to climate change mitigation and energy security are similar, policymakers must also be cognizant of the underappreciated differences between these goals and the risk that measures to address one goal could come at the expense of the other. While using pricing mechanisms is politically challenging, failure to impose modest and predictable price increases now will mean increased exposure to worse price shocks in the long run—just as a vaccine prevents a painful disease outbreak in the future by exposing people, in a controlled way, to low levels of a virus today. The environment, national security, and economic security of the United States will be served best by reframing the debate and focusing on long-run, sustainable affordability instead of on fleeting short-term goals.

Notes

1. Jacobe (2008); Fram (2008).
2. Pew Research Center for the People and the Press (2009).
3. Though only forty-eight senators voted for the legislation, six senators who were absent entered statements into the record saying that they would have voted for cloture had they been present.
4. IPCC (2007).
5. IPCC (1996).
6. Weitzman (2008).
7. Nixon (1974).
8. Roberts (2008).
9. EIA (2006b).
10. From August 2007 to August 2008, gasoline prices increased by $1.07 in the United States and by $1.20 in the United Kingdom (EIA 2008e).
11. Yergin (2006).
12. Barnes and others (2006).
13. EIA (2006a).
14. EIA (2008e).
15. EIA (2008b); EPRI (2006).
16. Worldwide natural gas consumption is projected to rise by more than 50 percent by 2030 (EIA 2008e). Hartley and Medlock (2006) predicts global natural gas prices could rise by 25 percent between 2008 and 2018. The EIA's *Annual Energy Outlook* reference case shows natural gas prices falling until about 2020, then rising to their current price by 2030. Its high-price case shows natural gas prices remaining steady until about 2020, then rising about 25 percent by 2030 (EIA 2008c).
17. EIA (2008e).
18. EIA (2008); Engelder and Lash (2008).
19. Ellerman (1995); Warell (2006).
20. Warell (2006); EIA (2008k).
21. Hamilton (1983); Hamilton and Herrera (2004).
22. Bernanke, Gertler, and Walston (1997).

23. Hamilton (2008).

24. Hamilton (2009).

25. Orszag, Rubin, and Sinai (2004).

26. CBO (2006); Nordhaus (2007).

27. Parry, Wells, and Harrington (2007); Leiby (2007)

28. Delucchi and Murphy (2008); International Center for Technology Assessment (1998).

29. The Energy Information Administration predicted that average regular gas prices would rise to $3.86 in the fourth quarter of 2008 but then decline to $3.73 by the end of 2009 (EIA 2008g). Economists at the Dallas Federal Reserve Board are even more skeptical that oil prices will remain at current levels, writing that "Absent supply disruptions, it will be difficult to sustain oil prices above $100 (in 2008 dollars) over the next 10 years" (Brown, Virmani and Alm 2008). By contrast, Goldman Sachs analysts in May 2008 predicted oil prices reaching between $150 and $200 per barrel by 2010 (Murti and others 2008). Similarly, economist Jeff Rubin at Canadian brokerage CIBC World Markets predicts oil prices reaching $200 per barrel by 2010 (Rubin 2008).

30. CERA (2008).

31. IEA (2008b).

32. Dargay, Gately, and Sommer (2006).

33. Chen and Jaffe (2007).

34. Sandalow (2007).

35. Sandalow (2008).

36. Sandalow (2008).

37. Author's calculations, based on Edwin M. Truman's May 21, 2008, testimony before the U.S. House of Representatives Committee on Foreign Affairs, table 1 (Truman 2008).

38. Truman (2007); Graham and Marchick (2006); Marchick and Slaughter (2008).

39. Friedman (2006); CFR (2006).

40. Collier (2008). For a contrary viewpoint, see Alexeev and Conrad (2008), in which the authors argue that the effect of a large endowment of oil and other mineral resources on long-term economic growth of countries has been on balance positive and that there is no correlation between a country's natural resource endowments and the quality of its institutions.

41. BEA (n.d.).

42. BLS (2006).

43. Clifford Krauss, "Rural U.S. Takes Worst Hit as Gas Tops $4 Average," *New York Times,* June 9, 2008.

44. EIA (2007).

45. CBO (2008a).

46. Hamilton (2009).

47. For the growth rate of Chinese emissions, see Netherlands Environmental Assessment Agency (2008). The cumulative CO_2 emissions in the 450 parts-per-million scenario are based on an average of the three models (MERGE, MiniCAM, and IGSM) used in Clarke and others (2007). Many thanks to Joe Aldy of Resources for the Future for sharing this analysis.

48. Burton, Diringer, and Smith (2006).

49. Energy Information Administration, "A Primer on Gasoline Prices" (www.eia. doe.gov/bookshelf/brochures/gasolinepricesprimer/).

50. Stavins (2007).

51. Stavins (2007); EIA (2008d).

52. IPCC (2005).

53. EPA (2007); DOE (2006).

54. Searchinger and others (2008).

55. EIA (2008b).

56. EIA (2008c, figure 3.1).

57. Point Carbon (2008). That trade-off may become less stark as an increasing share of natural gas consumed in the United States comes from liquefied natural gas, which, when the process of liquification and regassification is accounted for, is only 30 percent less carbon-intensive than coal (compared with 45 percent less for natural gas itself) (Jaramillo and others 2007).

58. Furman and others (2007).

59. Notably, energy prices for consumers will go up in a cap-and-trade system whether firms receive allowances for free or pay for them at auction. Even if allowances are auctioned, firms will still pass costs on to consumers because allowances can be sold for value in a liquid marketplace; therefore firms will want to recoup the opportunity cost of using an allowance (Bordoff 2008).

60. Stavins (2007).

61. CBO (2008a).

62. Delucchi and Murphy (2008); Leiby (2007).

63. Sandalow (2007); CFR (2006).

64. N. Gregory Mankiw, "Raise the Gas Tax," *Wall Street Journal*, October 20, 2006.

65. Indeed, many OECD countries may be taxing gasoline well beyond most estimates of the externality associated with oil consumption, although there may be other externalities that nations target through gas taxes, such as congestion or accidents. As noted in Deshpande and Elmendorf (2008), those are more efficiently addressed though targeted policies such as charges for vehicle miles traveled and congestion pricing. However, if the gasoline tax is the only available mechanism, those externalities should also be incorporated into the price of fuel.

66. In addition to gasoline tax proposals from those such as Mankiw, Martin Feldstein, former chairman of the CEA under President Reagan, has called for a system of tradable gasoline rights that would do for gasoline what cap and trade does for carbon (Martin Feldstein, "Tradeable Gasoline Rights," *Wall Street Journal*, June 5, 2006).

67. Sandalow (2007).

68. Once the tax reaches $40 per barrel (or whatever the externality of oil consumption is agreed to be), it should not be allowed to rise above that amount, even if the price of oil continues to decline. Unlike price-floor proposals that would prevent the price of oil from dropping below a certain amount (say $100 per barrel), the tax is not aimed at achieving a certain arbitrary price level, but rather making consumers bear the external cost of oil consumption. Once they have done so, there is little justification for preventing consumers from enjoying the benefits of more affordable oil if the market price declines significantly. Moreover, a tax of $40 per barrel is still high enough to provide developers of alternative fuels with some certainty that a market will exist for their products and to prevent OPEC from driving down prices temporarily by raising production.

69. Note that eventually the tax will almost inevitably hit its target level and will stay there, thus pricing the externality correctly regardless of the market price of oil.

70. Kahneman and Tversky (1979); Tversky and Kahneman (1991). Charles Schultze, former chairman of the Council of Economic Advisers, explained the concept of loss aversion in the context of budgetary policy: "The central implication of this view [loss aversion] for the politics of budgetary choices is that a tax increase—that is, an out-of-pocket loss of after-tax income—is viewed more negatively by voters than would be an agreement to forgo a tax cut of equal magnitude. Voters are very loath to support a general tax increase to raise federal civilian spending; but they will forgo a tax cut when defense spending declines as a way of reallocating the resources to civilian programs" (Schultze 1992, p. 38).

71. Leiby (2007).

72. EIA (2008h).

73. DOT (2008).

74. APTA (2008).

75. Brady and Gross (2008).

76. CBO (2008b).

77. DOE (2006).

78. Metcalf (2007).

79. Various policies to lower other costs associated with driving can also help ease the pain of higher gas prices. For example, switching from lump-sum to per-mile pricing of auto insurance would reduce annual auto insurance premiums for two-thirds of households, with an annual savings for those households of $270 per vehicle (Bordoff and Noel 2008). Because low-income people drive fewer miles on average, they would benefit especially.

80. Drilling in the Arctic National Wildlife Refuge, for example, would not meet such a threshold, as it would do little to reduce oil prices and would entail significant environmental risks. See EIA (2008a) and Kotchen and Burger (2007).

81. McPhail and Babcock (2008).

82. "Crude Measures," *The Economist*, May 29, 2008.

83. Governments often justify energy subsidies on the grounds that they help poor families afford energy. In some cases energy subsidies may increase energy affordability, but in other cases they can end up hurting the poor. For example, when subsidies create energy shortages, it is often the privileged or wealthy who get priority for energy rations. Moreover, direct cash transfers to families are often more beneficial than subsidizing a particular good, since families can spend the money in a way that maximizes their well-being. See United Nations Environment Programme (2008).

84. Andrew Batson and Neil King Jr., "China Lifts Fuel Prices and Oil Falls in Response," *Wall Street Journal*, June 20, 2008.

85. Brown, Virmani, and Alm (2008).

86. Ogden, Podesta, and Deutch (2007).

87. Metcalf (2008).

88. Searchinger and others (2008). While corn ethanol may yet turn out to be useful if it facilitates the transition to more climate- and food-friendly cellulosic ethanol, the manner of government involvement has certainly been suboptimal. If the government had instituted a 51 cent tax on oil instead of giving a 51 cent tax credit specifically to corn-based ethanol, then private industry would have had the incentive to develop a whole host of potential alternatives, not just corn ethanol. There is a strong rationale for the government to subsidize basic energy research instead of picking winners and losers, as discussed below.

89. Indeed, roughly 22 percent of the fuel savings from higher fuel efficiency is lost to such a "rebound" effect in the long run (Small and Van Dender 2007).

90. Although that claim is perhaps counterintuitive, a simple numerical example helps make the point. Consider two drivers: one increases fuel economy from 10 mpg to 12 mpg while the other switches from 50 mpg to 100 mpg. If they both drive 200 miles from Washington, D.C., to New York, the former will use 16.67 gallons rather than 20 gallons of gas (saving 3.33 gallons), while the latter will use 2 gallons rather than 4 gallons (saving just 2 gallons).

91. Bordoff (2009).

92. National Research Council (2001). But energy research subsidies, too, have diminishing returns. Newell (2008) argues that the federal climate R&D budget should be doubled, but beyond that extra spending begins to lose its social payoff. The lesson for policymakers is that only a small portion of the enormous revenue from a cap-and-trade system should be spent on clean technology research.

93. Newell (2008).

94. Darby (2006).

95. Darby (2006).

96. Michael S. Rosenwald, "For Hybrid Drivers, Every Trip Is a Race for Fuel Efficiency," *Washington Post*, May 26, 2008. The EPA estimates that driving less aggressively, for example, can increase fuel efficiency by 33 percent during highway driving (www.fueleconomy.gov/FEG/driveHabits.shtml).

References

Alexeev, Michael, and Robert Conrad. 2008. "The Elusive Curse of Oil." *Review of Economics and Statistics* 91, no. 3 (February), pp. 586–98.

American Public Transportation Association (APTA). 2008. Public Transportation Ridership Statistics (www.apta.com/research/stats/ridership/).

Barnes, Joe, and others. 2006. "Introduction to the Study." In *Natural Gas and Geopolitics from 1970 to 2040*, edited by David G. Victor, Amy M. Jaffee, and Mark H. Hays. Cambridge University Press.

Blanchard, Oliver J., and Jordi Gali. 2007. "The Macroeconomic Effects of Oil Shocks: Why Are the 2000s So Different from the 1970s?" Working Paper 13368. Cambridge, Mass.: National Bureau of Economic Research (September).

Bernanke, Ben, Mark Gertler, and Mark Walston. 1997. "Systematic Monetary Policy and the Effects of Oil Price Shocks." *BPEA*, no. 1, pp. 91–142.

Bordoff, Jason E. 2008. "International Trade Law and the Economics of Climate Policy: Evaluating the Legality and Effectiveness of Proposals to Address Competitiveness and Leakage Concerns." Brookings (June).

———. 2009. "Refuel Economy with Cash for Old Cars." Brookings (January).

Bordoff, Jason E., and Pascal J. Noel. 2008. "Pay-As-You-Drive Auto Insurance: A Simple Way to Reduce Driving Related Harms and Increase Equity." Hamilton Project Discussion Paper 2008-9. Brookings.

Brady, Aaron, and Samantha Gross. 2008. "Drivers Turn the Corner in the United States: Gasoline 'Peek Demand' Sooner than Expected?" Decision Brief. Cambridge Energy Research Associates (June 3).

Brown, Stephen P. A., Raghav Virmani, and Richard Alm. 2008. "Economic Letter: Insights from the Federal Reserve Bank of Dallas." *Federal Reserve Bank of Dallas* 2, no. 5 (May).

Bureau of Economic Analysis (BEA). n.d. National Income and Product Accounts. Table 2.3.5 "Personal Consumption Expenditures by Major Type of Product."

———. National Income and Product Accounts. Table 1.1.5 "Gross Domestic Product. Seasonally Adjusted at Annual Rates."

Bureau of Labor Statistics (BLS). 2006. "2006 Consumer Expenditure Survey." Washington.

Burton, Ian, Elliot Diringer, and Joel Smith. 2006. "Adaptation to Climate Change: International Policy Options." Paper prepared for the Pew Research Center on Climate Change (November).

CERA. 2008. "Capital Costs Analysis Forum-Upstream: Market Review." CERA Special Report.

Chen, Matthew E., and Amy Myers Jaffe. 2007. "Energy Security: Meeting the Growing Challenge of National Oil Companies." *Whitehead Journal of Diplomacy and International Relations.*

Clarke, Leon E., and others. 2007. "Scenarios of Greenhouse Gas Emissions and Atmospheric Concentrations." U.S. Climate Change Science Program, Synthesis and Assessment Product 2.1a (July).

Collier, Paul. 2008. *The Bottom Billion: Why the Poorest Countries are Failing and What Can be Done about It.* Oxford University Press.

Congressional Budget Office (CBO). 2006. "The Economic Effects of Recent Increases in Energy Prices" (July).

———. 2008a. "Policy Options for Reducing CO_2 Emissions" (February).

———. 2008b. "Effects of Gasoline Prices on Driving Behavior and Vehicle Markets" (January).

Council on Foreign Relations (CFR). 2006. "National Security Consequences of U.S. Oil Dependency." Independent Task Force, chaired by John Deutch and James Schlesinger. New York.

Darby, Sarah. 2006. "The Effectiveness of Feedback on Energy Consumption: A Review for DEFRA of the Literature on Metering, Billing, and Direct Displays." Environmental Change Institute, Oxford University (April).

Dargay, Joyce, Dermot Gately, and Martin Sommer. 2006. *Vehicle Ownership and Income Growth Worldwide: 1960–2003.* Leeds, U.K.: Institute for Transport Studies, University of Leeds.

Dell, Melissa, Benjamin F. Jones, and Benjamin A. Olken. 2008. "Climate Change and Economic Growth: Evidence from the Last Half-Century." Working Paper 14132. Cambridge, Mass.: National Bureau of Economic Research (June).

Delucchi, Mark A., and James Murphy. 2008. "U.S. Military Expenditures to Protect the Use of Persian Gulf Oil for Motor Vehicles." *Energy Policy* 36, no. 6.

Department of Energy (DOE). 2006. "Report to Congress on the Interdependency of Energy and Water" (December).

Department of Transportation (DOT). 2008. "Traffic Volume Trends" (www.fhwa.dot.gov/ohim/tvtw/tvtpage.htm).

Deshpande, Manasi, and Douglas Elmendorf. 2008. "An Economic Strategy for Investing in America's Infrastructure." Hamilton Project Strategy Paper. Brookings.

Deutch, John, and James R. Schlesinger. 2006. "National Security Consequences of U.S. Oil Dependency." Independent Task Force Report 58. Washington: Council on Foreign Relations.

Electric Power Research Institute (EPRI). 2006. "Generation Technologies for a Carbon Constrained World." *EPRI Journal* (Summer).

Ellerman, A. Denny. 1995. "The World Price of Coal." *Energy Policy* 23, no. 6, pp. 499–506.

Energy Information Administration (EIA). n.d. "Weekly Retail Gasoline and Diesel Prices." Washington: U.S. Department of Energy.

———. 2006a. "Electric Power Annual 2006" (November).

———. 2006b. "United Kingdom Country Analysis Brief" (May).

———. 2007. "Emissions of Greenhouse Gases Report: Carbon Dioxide Emissions" (November 28).

———. 2008a. "Analysis of Crude Oil Production in the Arctic National Wildlife Refuge" (May).

———. 2008b. "Annual Energy Outlook 2008" (June).

———. 2008c. "Annual Energy Review 2007" (June).

———. 2008d. "Energy Market and Economic Impacts of S. 2191, the Lieberman-Warner Climate Security Act of 2007" (April).

———. 2008e. "International Energy Outlook 2008" (June).

———. 2008f. "International Energy Price Information, Motor Gasoline" (www.eia.doe.gov/emeu/international/prices.html#Motor).

———. 2008g. "Monthly Energy Review" (July).

———. 2008h. "Short Term Energy Outlook" (August).

———. 2008i. "U.S. Product Supplied for Crude Oil and Petroleum Products." (http://tonto.eia.doe.gov/dnav/pet/pet_cons_psup_dc_nus_mbblpd_a.htm).

———. 2008j. "U.S. Crude Oil, Natural Gas, and Natural Gas Liquids Reserves: 2007 Annual Report." DOE/EIA-0216. (www.eia.doe.gov/pub/oil_gas/natural_gas/data_publications/advanced_summary/current/adsum.pdf).

———. 2008k. International Energy Annual 2006. Table 8.2 "World Estimated Recoverable Coal" (www.eia.doe.gov/pub/international/iea2003/table82.xls).

Engelder, Terry, and Gary G. Lash. 2008. "Marcellus Shale Play's Vast Resource Potential Creating Stir in Appalachia." *American Oil and Gas Reporter*. May.

Environmental Protection Agency (EPA). 2007. "Greenhouse Gas Impacts of Expanded Renewable and Alternative Fuel Use." 420-F-07-035. Office of Transportation and Air Quality (April).

Federal Reserve Bank of St. Louis. 2008. "Series: OILPRICE, Spot Oil Price: West Texas Intermediate" (http://research.stlouisfed.org/fred2/series/OILPRICE).

Fischer, Carolyn, and Richard G. Newell. 2007. "Environmental and Technology Policies for Climate Mitigation." Discussion Paper. Washington: Resources for the Future (February).

Fram, Alan. 2008. "9 in 10 See Rising Gas Prices Causing Family Hardship." Yahoo News (http://news.yahoo.com/page/election-2008-political-pulse-gas-prices).

Friedman, Thomas L. 2006. "The First Law of Petropolitics." *Foreign Policy* (May–June).

Furman, Jason, Jason E. Bordoff, Manasi Deshpande, and Pascal J. Noel. 2007. "An Economic Strategy to Address Climate Change and Promote Energy Security." Hamilton Project Strategy Paper. Brookings.

Graham, Edward M., and David M. Marchick. 2006. "U.S. National Security and Foreign Direct Investment." Washington: Peterson Institute for International Economics.

Greene, David L. 2008. "Measuring Energy Security: Can the United States Achieve Oil Independence?" Oak Ridge National Laboratory (May 8).

Gosselin, Peter. 2008. "High Wire: The Precarious Financial Lives of American Families."

Hamilton, James D. 1983. "Oil and the Macroeconomy since World War II." *Journal of Political Economy* (April), pp. 228–48.

———. 2008. "Oil and the Macroeconomy." In *The New Palgrave Dictionary of Economics*, edited by Steven N. Durlauf and Lawrence E. Blume. Palgrave Macmillan.

———. 2009. "Causes and Consequences of the Oil Shock of 2007–08." *BPEA*, no. 1.

Hamilton, James D., and Anna M. Herrera. 2004. "Oil Shocks and the Aggregate Macroeconomic Behavior: The Role of Monetary Policy." *Journal of Money, Credit, and Banking* 36 (April), pp. 265–386.

Hartley, Peter, and Kenneth B. Medlock. 2006. "The Baker Institute World Gas Trade Model." In *Natural Gas and Geopolitics from 1970 to 2040*, edited by David G. Victor, Amy M. Jaffe, and Mark H. Hays. Cambridge University Press.

Intergovernmental Panel on Climate Change (IPCC). 1996. "Climate Change 1995: Impacts, Adaptation, and Vulnerability." Contribution of Working Group II to the Second Assessment Report of the IPCC. Cambridge University Press.

———. 2005. "Carbon Dioxide Capture and Storage." Paper prepared by Working Group III. Cambridge University Press.

———. 2007. "Climate Change 2007: Synthesis Report." Geneva.

International Energy Agency (IEA). 2008a. "End-User Petroleum Product Prices and Average Crude Oil Import Costs" (July).

———. 2008b. "Oil Market Report" (July 10). (http://omrpublic.iea.org/current issues/full.pdf).

International Center for Technology Assessment. 1998. "The Real Price of Gasoline." Washington (November).

Jacobe, Dennis. 2008. "Economic Issues Reaching 'Crisis' Level for Many Americans." Gallup (May 1) (www.gallup.com/poll/106939/Economic-Issues-Reaching-Crisis-Level-Many-Americans.aspx).

Jaramillo, P., W. M. Griffin, and H. S. Matthews. 2007. "Comparative Life Cycle Air Emissions of Coal, Domestic Natural Gas, LNG, and SNG for Electricity Generation." *Environmental Science and Technology* 41, pp. 6290–296.

Kahneman, Daniel, and Amos Tversky. 1979. "Prospect Theory: An Analysis of Decision under Risk." *Econometrica* 47 (March), pp. 263–91.

Kotchen, Matthew J., and Nicholas E. Burger. 2007. "Should We Drill in the Arctic National Wildlife Refuge? An Economic Perspective." *Energy Policy* 35, pp. 4720–29.

Leiby, Paul N. 2007. "Estimating the Energy Security Benefits of Reduced U.S. Oil Imports." Paper prepared by Oak Ridge National Laboratory for the U.S. Department of Energy. Oak Ridge, Tenn. (February 28).

Marchick, David M., and Matthew J. Slaughter. 2008. "Global FDI Policy: Correcting a Protectionist Drift." Washington: Council on Foreign Relations (June).

Massachusetts Institute of Technology (MIT). 2003. "The Future of Nuclear Power." MIT Interdisciplinary Study (July).

———. 2007. "The Future of Coal." MIT Interdisciplinary Study (March).

McPhail, Lihong Lu, and Bruce A. Babcock. 2008. "Short-Run Price and Welfare Impacts of Federal Ethanol Policies." Working Paper 08-WP 468. Iowa State University, Center for Agricultural and Rural Development. (June).

Metcalf, Gilbert E. 2007. "A Proposal for a U.S. Carbon Tax Swap: An Equitable Tax Reform to Address Global Climate Change." Hamilton Project Discussion Paper 2007-12. Brookings.

———. 2008. "Using Tax Expenditures to Achieve Energy Policy Goals." Working Paper 13753. Cambridge, Mass.: National Bureau of Economic Research. (January).

Murti, Arjun N., Brian Singer, Kelvin Koh, and Michele della Vigan. 2008. "$100 Oil Reality, Part 2: Has the Super-Spike End Game Begun?" Global Investment Research, Goldman Sachs Group, New York (May 5).

National Research Council (NRC). 2001. *Energy Research at DOE: Was It Worth It?* Washington: National Academies Press.

Netherlands Environmental Energy Agency. 2008. "Global CO_2 Emissions: Increase Continued in 2007" (June 13).

Newell, Richard. 2008. "An Innovation Strategy for Climate Change Mitigation." Hamilton Project Discussion Paper. Brookings.

Nixon, Richard. 1974. "State of the Union Address" (January 30) (http://74. 125.45.104/search?q=cache:S1pjgXQ_lUsJ:www.thisnation.com/library/sotu/ 1974rn.html+Nixon+Energy+Independence&hl=en&ct=clnk&cd=3&gl=us).

Nordhaus, William D. 2007. "Who's Afraid of the Big Bad Oil Shock?" BPEA, no. 2.

Ogden, Peter, John Podesta, and John Deutch. 2007. "The United States Energy Innovation Initiative." Washington: Center for American Progress (October).

Orszag, Peter R., Robert E. Rubin, and Allen Sinai. 2004. "Sustained Budget Deficits: Longer-Run U.S. Economic Performance and the Risk of Financial and Fiscal Dissaray." Paper presented at the AEA-NAEFA Joint Sessions (January 5).

Parry, Ian W. H., Margaret Wells, and Winston Harrington. 2007. "Automobile Externalities and Policies." Discussion Paper 06-26. Washington: Resources for the Future.

Pew Research Center for the People and the Press. 2009. "Economy, Jobs Trump All Other Policy Priorities in 2009." Washington (January).

Point Carbon. 2008. "Carbon Market Europe, June 13" (www.pointcarbon.com/news/cme/1.934956).

Roberts, Paul. 2008. "The Seven Myths of Energy Independence." *Mother Jones* (May–June).

Rubin, Jeff. 2008. "Heading for the Exit Lane." New York: CIBC World Markets Inc. (June 26).

Sandalow, David. 2007. *Freedom from Oil.* New York: McGraw-Hill.

———. 2008. "Rising Oil Prices, Declining National Security." Testimony before House Committee on Foreign Affairs. May 29.

Schultze, Charles. 1992. "Is There a Bias toward Excess in U.S. Government Budgets or Deficits?" *Journal of Economic Perspectives* 6, no. 2 (Spring).

Searchinger, Timothy, and others. 2008. "Use of U.S. Croplands for Biofuels Increases Greenhouse Gases through Emissions from Land Use Change." *Science* 319, no. 5867 (February 29), p. 1238.

Small, Kenneth A., and Kurt Van Dender. 2007. "Fuel Efficiency and Motor Vehicle Travel: The Declining Rebound Effect." *Energy Journal* 28, no. 1.

Stavins, Robert N. 2007. "A U.S. Cap-and-Trade System to Address Global Climate Change." Hamilton Project Discussion Paper 2007-13. Brookings.

Truman, Edwin M. 2007. "Sovereign Wealth Funds: The Need for Greater Transparency and Accountability." Policy Brief in International Economics. Washington: Peterson Institute for International Economics (August).

———. 2008. "The Rise of Sovereign Wealth Funds: Impacts on U.S. Foreign Policy and Economic Interests." Testimony before the U.S. House of Representatives Committee on Foreign Affairs. May 21.

Tversky, Amos, and Daniel Kahneman. 1991. "Loss Aversion in Reckless Choice: A Reference Dependent Model." *Quarterly Journal of Economics* 106, no. 4 (November), pp. 1039–61.

United Nations Environment Programme. 2008. "Reforming Energy Subsidies: Opportunities to Contribute to the Climate Change Agenda" (www.unep.org/pdf/PressReleases/Reforming_Energy_Subsidies.pdf).

Warell, Linda. 2006. "Market Integration in the International Coal Industry: A Co-integration Approach." *Energy Journal* 27, no. 1, pp. 99–118.

Weitzman, Martin L. 2008. "On Modeling and Interpreting the Economics of Catastrophic Climate Change." Department of Economics, Harvard University (July).

Yergin, Daniel. 2006. "Ensuring Energy Security." *Foreign Affairs* (March–April).

CHAPTER TEN

Five "G's"

Lessons from World Trade
for Governing Global Climate

WILLIAM ANTHOLIS

Reversing the greenhouse gas (GHG) emissions of the world's $60 trillion economy will be among the most complex international governance challenges ever, rivaling the forty-year effort to dramatically reduce tariffs and establish a rules-based trading system. Given that nearly fifteen years have passed since the completion of the last global trade pact, it is easy to forget that the World Trade Organization (WTO) stands tall among the great successes of global governance precisely because it was able to accomplish what it set out to do despite the difficulties involved. A counterpart institution—a global system to address climate change—can be constructed that mimics the trade regime's most successful governance principles and avoids its structural weaknesses. Indeed, it would be both unfortunate and ironic if a global climate regime could succeed only at the expense of the global trade regime—or vice versa. What lessons should a climate regime learn from the trade regime? It may be helpful to break the issue down into five core questions: Who governs? What is the structure of the basic governing agreement? In what way is it "binding"? When can the agreement be expected to take effect? How does it bring in new nations? For each question, preliminary answers can be found in

The author is indebted to Scott Barrett, Colin Bradford, Lael Brainard, Daniel Drezner, Stuart Eizenstat, Alex Fife, Lauren Fine, Warwick McKibbin, Carlos Pascual, Nigel Purvis, David Sandalow, and Strobe Talbott for their helpful comments.

what can be thought of as the five "Gs" that should govern climate change. By looking to the lessons learned from the WTO, I try to make the case for a climate regime that

—starts with a *group* of major emitters, which together
—forge a *general agreement* to tackle the issue, one that
—*gears up* nations' domestic actions and
—organizes itself around a *generational* goal that
—allows for the *graduation* of developing countries into full commitment.

In a few of these areas, such an approach can provide a roadmap to resolving potential conflicts between the two regimes.

Who Governs? The Right *Group* of Nations, Matched to the Challenge

International regimes need to be designed to accomplish their purposes. Are they debating forums? Are they negotiated agreements that govern in particular fields? Trade and climate change both have benefited considerably from both kinds of organization. This chapter assumes that concerned nations are moving toward a governing regime for GHG emissions and that they need mechanisms that allow them to address that challenge.

Since the formation of the UN system, two bodies have existed alongside one another on the issue of global trade, one for discourse, the other for governance. The UN Conference on Trade and Development (UNCTAD) has functioned largely as a forum for assessing the twin goals and accomplishments of trade and development. Alongside UNCTAD, the General Agreement on Tariffs and Trade (GATT) and its successor, the World Trade Organization, have been the governing body for global trade. Though some might find it odd to point to the WTO as a successful model of international governance (especially given recent difficulties in completing the Doha round of multilateral negotiations), it is easy to forget how significant its contributions have been to both international cooperation and economic growth over the last sixty years.[1] The GATT/WTO system began when a group of the right countries decided to work together, as both a smaller (in terms of membership) and more ambitious (in terms of governance) world body than UNCTAD.

Lesson learned: size matters. When it comes to global governance, it was and is easier to get things done with a smaller number of the right countries. The GATT process was managed by the biggest and most technically competent trade players—the co-called Quad, composed of the United States, Japan, Canada, and the European Commission. Occasionally, when formal negotiations bogged down, the Group of Seven (and later Group of Eight) would weigh in to give the talks a boost, such as in 1978 and 2001, when the leaders themselves helped spur breakthroughs leading, respectively, to the close of the Tokyo round and the launch of the Doha round.

As the WTO's membership grew in size over its first five decades, negotiations became more unwieldy. The greatest number of new entrants came from developing countries. After an initial sorting out, the lesson of size was relearned: a new Quad was established, in which India and Brazil joined the United States and the EU as the principal negotiators. Further complicating matters, over the years a plethora of regional and bilateral agreements have advanced trade liberalization worldwide. The EU has led the pack in depth of integration and effectiveness, but the last forty years have seen the rise of a South American commercial union (Mercosur), the North American Free Trade Agreement, the Southern African Customs Union (SACU), and the Association of Southeast Asian Nations Free Trade Area. Of course, there is considerable debate about whether this spaghetti soup of different agreements has been good for the global trading system. Supporters of the three-way street (global, regional, and bilateral agreements) have found "competitive liberalization" to be a positive force. Regional agreements help drive reluctant countries to global negotiations for fear of missing gains from trade. Opponents see the growing complexity and difficulty of multiple trade talks to exceed the negotiating capacity of diplomats and the political will of elected officials. Complexity is unavoidable, to be sure. That the complexity has been manageable at all is due, in part, to the bedrock of a rules-based system that was established sixty years ago and the committed leadership of a relatively small number of players.

So what does that mean for the climate change regime? The half-true cliché about climate change is that it is a global problem that requires a global solution. Still, moving forward does not require all countries to be part of the solution, at least not at first. The UN-sponsored Kyoto Protocol process was slowed down by trying to conduct a global research initiative on the nature of the challenge—led largely by the UN's Intergovernmental Panel on Climate Change (IPCC)—while also debating who was responsible for addressing the challenge and negotiating an agreement among 140 nations under the UN's Framework Convention on Climate Change (UNFCCC). Though data, debate, and dialogue were critical to convincing those nations of the challenge at hand, the negotiations over what to do about it became rancorous and left many questions unanswered. They gave way to several more years of disputes over how to implement the agreement, a lengthy and unsuccessful ratification discussion in the United States, and uninspiring results on the ground, even from enthusiastic backers like the EU and Japan, which face an uphill battle in meeting their 2008–12 GHG emission targets. Meanwhile, the main developing country bloc is an eclectic group, including nations ranging from giant powerhouses such as Brazil, China, and India to small, poor, landlocked nations in Africa to small island nations. With the exception of the island countries—which literally could get washed away if there is no progress—most have been quite comfortable with the UN's penchant for discussion, so long as those discussions do not lead to binding obligations for their own economies.

In short, what the world has is a large problem coupled with a complicated, bureaucratic, and torpid negotiating mechanism. If size matters when setting up a governing regime, then the climate system needs to separate the broad and inclusive dialogue about the challenge from the more narrow and detailed challenge of negotiating an agreement. The latter task is best undertaken by a smaller group of nations.[2]

The great bulk of GHG emissions likely to spew into the atmosphere over the next three decades—and the economic and technical capacity to reverse course—can be found in fewer than two dozen countries. The creation of smaller groups—such as a major emitters group (E-8)—could help to address the challenges of climate change.[3] The United States, the European Union, China, Russia, Japan, and India are the top six emitters of GHGs. South Africa and Brazil rank 10th and 13th, respectively, but because they are key representatives of their regions, their contributions to addressing global warming are significant. That is especially true for Brazil, where protecting the Amazon is a major priority for storing carbon. The same logic lies behind the meeting of major emitters that President George W. Bush hosted in September 2007, which, added to my list of eight, include Canada (7th), South Korea (8th), Mexico (9th), Indonesia (12th), and Australia (15th). Together, these thirteen countries produce more than 80 percent of all GHGs.

Keeping the core group of negotiating nations small—and occasionally involving heads of state in the conversations—has one other signal virtue.

The same set of players is at the center of WTO negotiations. As the two regimes begin to bump into each other on a range of issues—from border surcharges to energy subsidies—resolution can be reached more easily if the same players from both regimes are talking. That is especially true if heads of state themselves are aware of both the need to coordinate and the perils of failure to do so.

What is the Form of Governance? A *General Agreement*

One of the keys to the GATT/WTO's success is that it did not start as a global body but as a less formal arrangement. If that distinction seems unimportant, keep in mind that the WTO started not as the successful WTO—or even the successful GATT—but as the failed International Trade Organization (ITO), which was envisioned at Bretton Woods along with the World Bank and the International Monetary Fund. The treaty establishing ITO died on the floor of the U.S. Senate because two-thirds of that august body was not prepared to hand over highly political decisions regarding trade policy to an international organization. The negotiators went back to the drawing board. Only after the International Trade Organization's high-profile failure did they come up with the General Agreement on Tariffs and Trade.

The core lesson: do not start with an international treaty organization responsible for data, debate, and enforcement of the terms of the treaty. And when it comes to enforcement, build confidence through general agreements, which are binding only in that they synchronize and increase the ambition of domestic actions that states see as being in their best interest. For nearly fifty years, GATT was able to negotiate and adjudicate agreements that bound nations in a way that did not directly call national sovereignty into question. Each participating nation pledged to cut tariffs and other trade barriers in a coordinated way. Countries could choose what they thought counted as a significant reduction, and often they would trade fast action in one area for slow action in another. Once commitments were made, they had to be enforced. An adjudicative body was established to resolve trade disputes.

Technically speaking, the adjudicative trade body did not enforce the treaty. Member nations did. Countries monitored one another's behavior, including that of the most economically powerful trading nations. When a country had a complaint, it brought it to the GATT/WTO's dispute resolution body. If a defendant country lost a dispute, it had a choice: change its domestic law or allow a retaliatory tariff or other action by the plaintiff country. In that way, all countries felt the system to be self-enforcing. All of this gave negotiators the ability to say convincingly to their political masters—including their general publics—that the agreement did not sacrifice national sovereignty.

The fear that nations will lose their sovereignty has also plagued the climate change discussions. If the United States had ratified the Kyoto Protocol, it would have been a binding treaty. Opponents of Kyoto claimed that the United States would have been subject to some set of sanctions that would be administered and enforced by the UN. The nation's sovereignty over its energy future—and by extension, its national security—would be subject to external intervention. As a political matter, few U.S. politicians want to be told that they must do something or else face sanction by a global body.

Whether those concerns have any factual merit, "sovereignty hawk" nations around the world (particularly the United States and countries in much of the developing world) have feared Kyoto-style obligations. Political leaders in the United States, China, India, and Brazil have refused to sacrifice their ability to control their economic destinies to a global energy regime—or, at least, they have refused to give up their sovereignty in a way that diverges from their national interest. Only the European Union—whose members have grown comfortable sharing or even pooling their sovereignty—seems to like the idea of using an international agreement to compel domestic action.

There is another way, of course. Building on the successful GATT model, negotiators could seek a General Agreement to Reduce Emissions (GARE). Like GATT, the proposed GARE would effectively link domestic action with an international agreement.[4] It would avoid moving too quickly to a full-blown international institution, such as a World Environment Organization. If a "treaty" suggests that nations are tying their fates to one another, "general agreement" suggests that nations acknowledge one another's interdependence but also their autonomy. As they build confidence in their ability to work together under such an agreement, they may become more willing to strengthen the regime. A GARE system could be built on the E-8—the major emitters group outlined above. A core set of the most important countries could start the process, and this process ultimately would be compatible with regional and bilateral agreements. Each year, leaders of the group could meet to evaluate progress and give a boost to the ongoing negotiations.

What then of the UN? An important role remains for the UN in continuing to sponsor the broader climate talks as a forum for helping nations to share information and best practices with one another. The UN also has been path-breaking in supporting the critical work of the Intergovernmental Panel on Climate Change—the scientific body that has helped establish that climate change is real and that human action is contributing dramatically to it. Both functions help support the negotiation and conflict resolution functions of a binding agreement. Eventually, once confidence is built in a self-enforcing agreement, the UN can be brought in to maintain the relationships.

Where Does it Bind Nations? It *Gears Up* Domestic Actions that Nations are Willing to Take

Ask a State Department lawyer, and he or she will tell you that there is no difference between a treaty, a congressional-executive agreement, and a presidential bilateral statement issued with a foreign head of state. The United States is honor bound to live up to its agreements, whatever form they take. If the agreement includes consequences for violation, the United States is obligated to accept them. Yet in practice, nations (including the United States) frequently violate or ignore agreements—and suffer the consequences or not. Though the UN Charter provides some instances in which states may be physically compelled to act in accord with international norms, in practice that rarely is the case with non-military agreements.

What makes some international agreements binding? What makes some "bindings" succeed and others fail? There are at least three ways to view the success of binding agreements. First, some pacts succeed because states feel no need to violate them. Such agreements succeed because they create a structure that allows states to do what they would prefer to do but might not do because they fear noncompliance by others. By giving states the confidence that other states will live up to their end of the bargain, agreements allow states to do what is in their best interest. That is what de Tocqueville called "self-interest rightly understood."

Second, some agreements succeed because nations realize upon violating an agreement that the net costs of doing so are not worth it. That is usually the case when nations contemplate sanctions from an agreement—and the political impact that those sanctions could have domestically and internationally—and choose to get right with the law. Third, agreements work when nations suffer consequences for their violations and both the violating nation and the nation that applies the sanction feels that the consequences are appropriate and adequate.

In theory, none of the three cases requires an outside enforcing body. It is governance without government—or what the great international relations theorist Hedley Bull called "the efficacy of international law," which "depend[s] on measures of self-help."[5] The GATT/WTO succeeded because, for its first fifty years, all three forms of self-help worked. First, the commitments were sufficiently robust that countries could plan to cut trade barriers—that is, gear up their commitment—knowing that their counterparts would do the same. GATT/WTO negotiations helped nations to cut their own trade barriers further than they otherwise would have; in return, counterpart nations also lowered their barriers. Consumers benefited from cheaper imports, and exporters benefited from wider markets. Nations understood the tough domestic challenges other nations felt in trying to lower trade barriers.

This type of reciprocal action worked in practice, particularly when Congress signaled its willingness to lower barriers in specific product areas in advance of a negotiation. Making a priority of domestic action is actually enshrined in the domestic legal architecture of U.S. trade diplomacy. One reason that the United States is more easily bound by trade negotiations is that it uses congressional-executive agreements, which require passing relatively detailed trade promotion authority in advance of negotiations. As a result, the trading system aspired to adopt laissez-faire goals as a general matter across national boundaries but also accepted that national legislation was central to moving forward. Though laissez-faire remained a long-term goal, no single round or negotiation ever proposed to complete the process and each successive round depended on national action. The system recognized the domestic political and economic constraints that nations face in moving toward a globally integrated goal.[6]

Second, GATT's enforcement system sustained national cuts without appearing to undermine sovereignty. When a nation was found to be in violation of a trade rule, it had a choice: change its trade practice or accept reciprocal trade sanctions on other goods. Even under trying circumstances, nations were willing to go back and change domestic law in order to come into compliance. In such instances, countries have avoided the imposition of sanctions, and they have been unwilling to sustain extended tit-for-tat sanctions. Third, in those few cases in which sanctions have been applied, nations have generally been willing to accept them without imposing counter-sanctions. Rather than starting trade wars, the GATT/WTO system has prevented them. A similar logic can guide a GARE: countries can choose to cut their domestic GHG emissions in the way that makes most sense given their domestic constraints. Rather than make adhering to a treaty a goal in and of itself, a GARE would start with domestic legislation and help nations strengthen—that is, gear up—their ambition.

Nearly all nations recognize that cleaner energy production and protection of forests are worthwhile goals in themselves and that they should act to prevent irreversible climate change, and almost all nations have taken some steps in that regard. A diversity of approaches is appropriate. Countries use energy and regulate pollution very differently, and they also differ widely in their capacity to track emissions and enforce compliance. The United States and China, for instance, are especially dependent on carbon-intensive industries such as coal. Brazil, conversely, has huge sources of renewable resources, such as hydropower and biofuels, but it also is struggling to save its rain forest—one of the great carbon sinks in the world. It is clear that a one-size-fits-all approach will not work.

The threefold challenge for international negotiators is, first, how to get countries to take reciprocal domestic actions; second, how to structure compliance so that it reinforces or returns states to mutual action; and third, how to establish sanctions that nations can choose to accept as appropriate. Therefore, first, a GARE should begin with domestic action and use the negotiating process to gear up the ambition of states. States are "bound" to follow through on actions that they take on their own.[7] One way to make sure that that is the case is to legislate first and negotiate later. In the U.S. context, GARE would take advantage of congressional-executive agreements and avoid the treaty process. In a GARE, the domestic political hurdle to passage is whether to pass and implement domestic law. With the framework of such a domestic law in place, the international negotiations can focus on the level of ambition to which each country commits, in order to help ratchet up their actions. The diplomatic challenge becomes whether that level of commitment is acceptable to counterpart nations.[8]

This is in slight, but significant, contrast to the Kyoto Protocol's approach, which binds a state to the decisions of an international organization. [9] For instance, in the United States, ratifying a treaty not only requires a supermajority in one house of Congress, it also requires passage of implementing legislation in both houses. Agreements, by contrast, require majorities in both houses—first for authorization to negotiate, second for the final agreement itself. The authorization to negotiate—the so-called "fast track" in trade talks—gives negotiators a roadmap for what can be negotiated and as a result begins to involve members of Congress themselves in the talks.[10] In a real sense, for the United States a GARE would start with domestic action and seek to ratchet it upward, in sync with the actions of other nations.

Second, a GARE would need to be "binding" by addressing noncompliance. As with the early GATT system, it should include avenues for self-enforcement of the terms of the agreement by the nations that are party to it. Exactly how nations will self-enforce an agreement is still being debated. Some analysts have called for a common global carbon tax. Others have called for a "pledge and review" process, through which nations pledge to reduce GHGs and then review one another's progress on a regular basis. There may be merit to both kinds of agreements. Yet neither, on its face, appears to encourage the gearing up of domestic commitments while discouraging nations from breaking those commitments by imposing sanctions that deny nations the benefits of the agreement.[11]

One approach, in theory, does accomplish those goals: international trading of GHG emissions. As a domestic matter, the EU has already adopted emissions trading, and the United States is considering such legislation, having successfully pioneered a sulfur dioxide system under President George H. W. Bush in the late 1980s. Though there have been some initial problems with the EU's system, it has now done largely what it was intended to do: put a price on carbon emissions and create incentives for the private sector to find emissions cuts where it is most efficient to do so.

International emissions trading would extend these advantages across national borders. The United States insisted on GHG emissions trading at Kyoto, and for nearly two years afterward it haggled with the European Union over the rules. Ironically enough, once the United States walked away from emissions trading, during the George W. Bush presidency, the EU began to pursue international emissions trading aggressively. Trading can happen in two forms—in a closed system or an open system. In a closed system, two different national economies agree that total emissions in both economies will not exceed a fixed amount; as long as both nations comply in the aggregate, permits would remain of equal value and be freely tradable between countries. If one country violates its emissions limits, however, the permits in that country become less valuable. In an open system, nations are responsible only for their own reductions, though investors or companies may seek certifiable reductions in other countries and may simply be free to invest in such reductions.[12]

Both approaches have strengths and weaknesses from the standpoint of "compliance as self-help." The strength of the closed system is that it raises the stakes for compliance—and the penalties for noncompliance. In such a system, it is highly advantageous for nations to make broad progress on their GARE reduction commitments, as it would either force nations to seek permits from firms that have successfully cut GHG emissions in other nations or provide incentives for nations to have the most number of such firms in their own territory. If such a system can be set up, the incentives for success should be high. Yet the cost of failure should also be high, as less successful countries would be forced to pay dearly for emissions permits across borders. In contrast, an open system would create incentives for investing across borders. That said, it would entail few downsides if nations failed to comply with the international agreement, other than the greater risk of failing to stabilize the climate.

The joint challenges for a GARE that relied on trading for compliance would be to determine whether a member country seeking to join had proposed a strong enough target and whether existing members had come close enough to their previous commitments in each successive round of negotiations. The first task would need to fall to member states; the second task could fall to a joint review panel established by GARE countries. If a country failed to meet its target by reducing its emissions or by buying permits, it would forfeit the right to continue in GARE in future periods.[13]

Third, establishing a successful binding agreement requires addressing how to deal with those who refuse to join. A growing chorus is raising the idea of using actual trade protections—such as requiring a firm importing goods from countries that have not adopted sufficient emission reductions to purchase emissions permits equivalent to the carbon emitted in the production of those goods. The idea first arose in countries such as France, directed at the United States for not joining the Kyoto Protocol. Now that the United States is contemplating joining a post-Kyoto system, it is considering applying the same approach to developing countries that do not adopt binding targets. These "border permits" would be a way of imposing some sanction on nations that refuse to join or to comply with an emissions agreement—and thereby help distribute the costs of compliance. That has the potential to be a constructive way to think through the problem, but it also could undermine the trade regime, the climate regime, or both. The constructive element of such an approach would be to provide real leverage for nations to actually transfer the costs of noncompliance in an effort to address a global public good—something for which the trade regime allows exemptions.

The potentially disruptive element is that all nations do not recognize the public good, let alone the means to address it, in the same way. Developing countries, which likely would be the targets of such a system, would be almost certain to claim that border permits violate the WTO's rules against nondiscrimination and that a border permit provision does not meet the standard for getting an environmental exemption from those rules. Developing countries would likely argue that the "global public good standard" was not met because the current international climate treaty already embodies how the international community defines the climate challenge. That treaty, they would claim, explicitly demands that industrial nations act first and exempts developing countries from binding targets. Because the standing global consensus is that industrial nations must act first, any effort to use the trade regime to shift that burden would be seen as illegitimate.

So if industrial countries persisted in imposing such tariffs in order to build a more effective climate regime, they might undermine the WTO, regardless of how the dispute was settled. If a developing country claimed that border permits are a violation of WTO rules but lost the dispute, the victory for industrial countries would come as an additional blow to developing nations, on the heels of the WTO's long-stalled Doha round, which has failed to produce openings to industrial markets for products from developing countries. Conversely, a victory for developing countries might further undermine public support for the WTO within industrial nations, which continues to wane. Likewise, the effect on the climate regime could be enervating. Emerging market players such as Brazil, China, and India might feel that they were being forced into a climate agreement by being denied access to an international trading regime that they had worked hard to enter as full participants. And industrial countries might be less inclined to join the climate regime if border adjustments were found to be illegal vis-à-vis the WTO, because they might feel that their competitiveness would be further eroded.

Avoiding such a clash of global governance regimes should be a priority not only for leading nations but also for the heads of both global regimes. Avoiding a clash is perhaps the best argument for not treating these issues in isolation from one another or from broader global economic developments. Indeed, one of the ironies of the spread of democracy has been that democratic governments have to work so hard to pass domestic regulation that they are often the least inclined to take direction from international organizations. The relatively fragile support for international regimes should not be casually challenged, particularly in the name of establishing other regimes.

When Can We Expect the New Climate Regime to Take Effect? Over a *Generation*

The idea of extending the enforcement of commitments over time gets at a central element of any governance challenge. One of the great successes of the trade regime was that it built itself gradually—only after forty-five years of operating did it lead to a treaty organization. The long-term nature of the climate challenge means that solutions also must be long term. Today's warmer climate is the result of the accumulation of GHG emissions over the last half-century. Today's emissions add to those historic concentrations, and they already are locking in warmer temperatures well past the middle of this century. Little can be done now to stop that warming from happening. The effort to slow emissions over the next several decades therefore will have the most effect on temperatures in the second half of this century.

What is the appropriate long-term goal? The starting point for all climate negotiations, the 1992 Rio Treaty (ratified by the U.S. Senate and adopted worldwide), included an abstract long-term goal: "stabilization of greenhouse gas concentrations in the atmosphere at a level that would prevent dangerous anthropogenic interference with the climate system." The Kyoto Protocol was a practical attempt to implement the Rio Treaty, yet it set only one target—a short-term reduction of GHG emissions by industrial nations. That was seen as a first step toward the longer-term goal. But because it lacked any second or third step, it was widely criticized for not getting at the longer-term challenges.

As with the trade regime, the climate regime should keep the long-term focus that was part of Rio's plan, and it should be geared around a portfolio of long-term targets, including for concentration levels and global temperature change. As with any law or diplomatic agreement, those targets could be adjusted later as scientific and economic evidence was collected. But the key is to get some agreement on the long-term goals so that short-term steps can be seen in their broader context.

Right now, many scientists believe that dangerous interference with the climate could be avoided if temperature increase is limited to 2 degrees centigrade. Consensus estimates predict that doing so requires at least stabilizing GHG concentration levels at 550 parts per million (ppm) by 2050. If the E-8 adopted 2°C and 550 ppm as global goals and urged other nations to do the same, countries could then cut their short-term and long-term emissions to levels that they felt constituted effective and fair steps toward meeting those goals. When diplomats tried to negotiate over relatively short-term emissions cuts, they would be better able to explain to their political leaders and publics how each short-term target contributes to a longer-term effort. (Indeed, in the recent proposed Lieberman-Warner climate legislation, a series of emissions cuts are written in, extending to 2050.) As nations reached their shorter-term benchmarks, they could assess how they were doing in meeting that longer-term goal. Among other things, that would help industrial countries signal to developing countries what they consider to be a fair share of the burden for all nations over a future term and show that it is possible to achieve those marks without hurting economic growth.

Setting targets for temperature increase and gas concentrations can also help politicians, the media, and the public stay focused on the purpose of the undertaking: to cut emissions sufficiently to slow and eventually stop global warming. Though scientists now overwhelmingly agree that human activities are contributing to global warming, new evidence is coming in constantly. The consensus is being affirmed but also challenged, and evidence is updated on a nearly daily basis—resulting mostly in more dire warnings. Some scientists, for instance, now think that stabilization at 450 ppm is needed to prevent 2 degrees of warming. Of greater concern, 2°C of warming may not be so safe. Recent research, for instance, finds that the current level of warming is melting the Arctic ice cap faster than had been anticipated, potentially weakening its ability to reflect sunlight and cool the planet. If the ice cap were to disappear with less than 2°C of warming, that could be a tipping point that would lead to a more dramatic and dangerous shock to the Earth's climate.

How Does It Bring New Nations Into the Agreement?
It Must Provide a Path for *Graduation*

Perhaps the greatest lesson that the climate regime can learn from the trade regime is something that the latter has failed, so far, to entirely address: how to bring developing countries into the regime in a way that acknowledges their development challenges but also allows them to graduate to full responsibility as their economies grow.

The trading regime is now in the midst of its longest negotiating round in its sixty-year history, the so-called WTO Doha development round. One of the main reasons why concluding this round has been so difficult is that it is trying to address the regime's core weakness: that the two basic groups—the industrial countries and the developing countries—have differing sets of obligations. The developing countries enjoy "special and differential treatment," which means that they are exempt from the more drastic tariff reductions taken by industrial nations. The regime is asymmetrical, and it also is unclear how any developing nation would graduate to taking on an industrial-strength obligation, when the time was right. Thus, although the addition of developing countries has been critical to achieving global scope for the organization, it also has added to the complexity of the process—and to the current stalemate in negotiations.

As with the global trading system, the developing countries will ultimately need to graduate and become part of the post–Kyoto Protocol climate system. Kyoto was problematic in several regards, but perhaps its biggest drawback was that developing countries did not commit to cutting their GHG emissions; in fact, the treaty prevents them from establishing a binding target even if they choose to do so. Argentina, for instance, tried to take on a binding target in 1998, but it was prevented from doing so by other developing countries.

It certainly makes sense for developing countries to have different obligations, or obligations that kick in later, given the industrial world's historic responsibility for greenhouse gas emissions and its much greater wealth, along with the generational nature of the problem. But there is simply no way to solve the climate problem without the active involvement of the developing countries, which, according to current projections, will account for more than 70 percent of GHG growth in the next twenty-five years. Yet those countries show no willingness to accept Kyoto-style targets.

This catch-22 is not just a political problem; it is an economic issue that goes to the heart of getting clean energy markets up and running. Most industrial countries are now poised to take near-term and middle-term efforts to cut GHG emissions, which is already leading to some increased investment in clean energy. However, if the world economy is going to cut its carbon emissions by as much as 80 percent, enormous amounts of capital investment will be required to find transformative, carbon-free sources of energy. The more certain investors feel that the industrial countries will keep seeking ever deeper reductions in GHG emissions, the more likely they will be to commit that kind of capital up front. The key is for the industrial countries to signal their long-term cuts. But they are less likely to do so as long as developing country action is not a sure thing. Right now, the developing countries are saying that they will not act, and they are refusing to address the long-term challenge.

How can the international community break out of this box? The effort must begin with the industrial world, by responding realistically to developing countries' concerns about equity. The developing countries rightfully feel that the rich countries are largely responsible for the problem to date and probably for the global warming that will take place over the next fifty years. The industrial countries should not dismiss their concerns, especially because despite their recent economic gains, the developing countries, particularly China and India, still have a nearly unfathomable number of their citizens living in extreme poverty—well over a billion people combined in those two countries alone. In addition to taking seriously efforts to estimate how much the industrial countries have contributed to current GHG concentration levels, industrial nations should also consider very long-term targets that include consideration of per-capita emissions.

Second, the industrial countries should appeal to the developing countries' self-interest. Climate change is most likely to hurt poor countries the worst by accentuating droughts and severe storms, for which these nations are the least prepared. Moreover, many of these countries are facing the local air pollution that comes in the early stages of industrialization and the health effects of local air and water pollution, which could be lessened by early adoption of clean energy technology. Moreover, investing in energy efficiency and clean energy is ultimately cost-effective. One possible motive for joining a GARE would be the potential to earn emissions trading credits on a sizable scale. In the near term, that would mean continuing to explore opportunities to earn emissions reduction credits on a project-by-project basis. That could potentially build support within the developing countries for adopting countrywide emissions policies, linked to GARE.

And last, the industrial countries should not be shy about public diplomacy on climate change. Right now, the developing countries do not feel any public pressure to respond to climate change—which is probably not surprising, given the development challenges that many of them are facing. A public diplomacy strategy therefore is needed that stresses each topic noted above, from equity to self-interest to the power of global markets to help transfer technology and capital to developing countries. Of course, all of those efforts require that the real first steps be taken in the industrial world.

Conclusion

The political will to mount a global effort to reduce GHG emissions has begun to develop in the United States and even in a few key developing countries. Such support, however, is far from the dramatic shift in attitudes needed to establish a full-blown global institution to address the climate challenge. In addition to the costs associated with acting, a core concern is a familiar one in global governance: loss of sovereignty. There are some good reasons for that. Even for the most committed nations, the climate change challenge is of such great economic and environmental complexity that few politicians are likely to simply turn over the keys to their national policymaking to an international treaty organization.

In taking the first steps toward a global climate regime, the industrial nations can learn from the experience of the global trading regime in building participants' confidence in a self-regulating system. The GATT/WTO system was built on a small group of states that, through a general agreement, were able to gear up domestic action over a generation. The advantage of that approach is that it does not pose a direct challenge to national sovereignty. Instead, it coordinates the work of states in a way that respects the diversity of local governance, and therefore it has a greater chance of getting buy-in from key players. The challenges of such an approach are that it does not guarantee fast domestic action, that many smaller states will feel left out of the process, and that the transition to the system may be difficult for many participating states. Last, as with the trade regime, it must overcome the biggest challenge for global governance in today's world: how to graduate nations when they emerge from the development process into the industrial world.

Notes

1. John Ruggie, "International Regimes, Transactions, and Change: Embedded Liberalism in the Postwar Economic Order," in *International Regimes*, edited by

Stephen Krasner (Cornell University Press, 1983), pp. 195–231. That success was apparent twenty-five years ago, when the GATT system was held up as the model for global governance, including among "realist" theorists of international relations, who tend to hold a dim view of the efficacy of international institutions. While Ruggie would not classify himself as a realist, his general argument was accepted by realists such as Krasner. In the real world of politics, the acceptance of GATT and the WTO among U.S. political conservatives—including their willingness to accept binding decisions by international tribunals—is striking.

2. Charles P. Kindleberger, "International Public Goods without International Government," *American Economic Review* 1, no. 76 (1986), p. 2–13. One commentator questioned whether the problem of protecting the Earth's climate is analogous to that of expanding free trade. As a general matter, most analysts would agree that protecting the climate is a non-excludable public good, while free trade has been less so, since only the participants in a trading regime enjoy the benefits. Some might even question whether free trade is a public good. Indeed, a strong argument can be made that both a climate regime and a trade regime offer both excludable and non-excludable public goods. In trade, the excludable public goods are the reduced tariffs and lower trade barriers offered to members of the regime; the non-excludable good is a stable international economic order, which has economic and political benefits for all countries. In climate change, the non-excludable good would be climate protection; the excludable good would be an emissions trading regime. Many thanks to Lael Brainard for helping clarify this distinction.

3. Todd Stern and William Antholis, "Creating an E-8," *American Interest* 2 (January 2007), pp. 43–48.

4. See the first suggestion for such an approach in Todd Stern and William Antholis, "A Changing Climate: The Road Ahead for the United States," *Washington Quarterly* 31 (Winter 2007–08), pp. 175–88.

5. Hedley Bull, *The Anarchical Society* (Columbia University Press, 1977), p. 131 and chapter 6 generally.

6. Ruggie, "International Regimes, Transactions, and Change."

7. For an overview of what a domestic and international approach for the United States might look like, see Stern and Antholis, "A Changing Climate."

8. One model example for this would be the EU's proposal to unilaterally cut its emissions by 20 percent below 1990 levels in the post-Kyoto commitment period and to extend those cuts to 30 percent if an international agreement is reached.

9. If GARE was not a treaty, one advantage would be that it would avoid another major drawback of Kyoto: it would not need a two-thirds majority in the U.S. Senate, a minefield where countless treaties have died. Indeed, by the treaty process, internationally agreed emissions targets and timetables and the policies and regulations needed to comply with them become deeply enshrined in domestic law because they have been passed by a supermajority in the Senate. By contrast, GARE would require only simple majorities in both the House and the Senate, putting the domestic legislation horse in front of the global treaty cart—just the way it should be. See both Stern and Antholis, "A Changing Climate" and Nigel Purvis, Climate Advisors, "Treat Climate Like Trade: The Case for Climate Protection Authority," unpublished policy brief manuscript.

10. William Antholis and Strobe Talbott, "Tackling Trade and Climate Change: Leadership on the Home Front of Foreign Policy," in *Opportunity 08*, edited by Michael O'Hanlon (Brookings, 2007), pp. 63–67.

11. Jonathan Wiener, "Incentives and Meta-Architecture," in *Architectures for Agreement: Addressing Global Climate Change in a Post-Kyoto World*, edited by Joseph Aldy and Robert Stavins (Cambridge University Press, 2007), pp. 74–76.

12. As mentioned earlier, establishing an emissions trading system would move from the non-excludable public good system of climate protection to a system with excludable benefits: access to trading with other parties, with the enhanced efficiency and reduced compliance costs that that implies.

13. Stern and Antholis, "A Changing Climate," p. 183.

Contributors

William Antholis
Managing Director, Brookings Institution

Jason Bordoff
Associate Director for Climate Change, White House Council on Environmental Quality
Former Policy Director, The Hamilton Project, Brookings Institution

Marilyn A. Brown
Professor, School of Public Policy, Georgia Institute of Technology

Erin E. R. Carter
Former Program and Special Projects Manager, Governance Studies, Brookings Institution

Manasi Deshpande
National Economic Council
Former Senior Research Assistant, The Hamilton Project, Brookings Institution

Erica S. Downs
China Energy Fellow, John L. Thornton China Center, Foreign Policy Studies, Brookings Institution

Jonathan Elkind
Principal Deputy Assistant Secretary of Energy for Policy and International
   Energy, United States Department of Energy
Former Senior Fellow, Foreign Policy Studies, Brookings Institution

Ann Florini
Senior Fellow, Foreign Policy Studies, Brookings Institution

Suzanne Maloney
Senior Fellow, Saban Center for Middle East Policy, Foreign Policy Studies,
   Brookings Institution

Pietro Nivola
Douglas Dillon Chair, Senior Fellow, Governance Studies, and Codirector, Red
   and Blue Nation, Brookings Institution

Pascal Noel
Policy Adviser, National Economic Council
Former Research Analyst, The Hamilton Project, Brookings Institution

Michael O'Hanlon
Sydney Stein, Jr. Chair, Director of Research, and Senior Fellow, Foreign Policy
   Studies, and Director of Research, 21st Century Defense Initiative, Brookings
   Institution

Carlos Pascual
Ambassador of the United States to Mexico, United States Department of State
Former Vice President and Director, Foreign Policy Studies, Brookings Institution

Andrea Sarzynski
Assistant Research Professor of Public Policy, Trachtenberg School of Public
   Policy and Public Administration, George Washington University
Former Senior Research Analyst, Metropolitan Policy Program, Brookings
   Institution

Frank Southworth
Principal Research Scientist, School of Civil and Environmental Engineering,
   Georgia Institute of Technology

Evie Zambetakis
Senior Research Assistant and Research Coordinator, Energy Security Initiative,
   Foreign Policy Studies, Brookings Institution

Index

CPSIA information can be obtained at www.ICGtesting.com
Printed in the USA
BVOW070929210113

311073BV00002B/3/P